Digital Darwinism

Ralf T. Kreutzer • Karl-Heinz Land

Digital Darwinism
Branding and Business Models
in Jeopardy

 Springer

Copernicus Books is a brand of Springer

Ralf T. Kreutzer
Berlin School of Economics and Law
Berlin
Germany

Karl-Heinz Land
neuland GmbH & CoKG
Cologne
Germany

Translation from German language edition: *Digitaler Darwinismus* by Ralf T. Kreutzer and Karl-Heinz Land
© 2013 Springer Fachmedien Wiesbaden
Springer Fachmedien Wiesbaden is a part of Springer Science+Business Media
All Rights Reserved

ISBN 978-3-642-54400-2   ISBN 978-3-642-54401-9 (eBook)
DOI 10.1007/978-3-642-54401-9
Springer Heidelberg New York Dordrecht London

Library of Congress Control Number: 2014950109

© Springer-Verlag Berlin Heidelberg 2015
This work is subject to copyright. All rights are reserved by the Publisher, whether the whole or part of the material is concerned, specifically the rights of translation, reprinting, reuse of illustrations, recitation, broadcasting, reproduction on microfilms or in any other physical way, and transmission or information storage and retrieval, electronic adaptation, computer software, or by similar or dissimilar methodology now known or hereafter developed. Exempted from this legal reservation are brief excerpts in connection with reviews or scholarly analysis or material supplied specifically for the purpose of being entered and executed on a computer system, for exclusive use by the purchaser of the work. Duplication of this publication or parts thereof is permitted only under the provisions of the Copyright Law of the Publisher's location, in its current version, and permission for use must always be obtained from Springer. Permissions for use may be obtained through RightsLink at the Copyright Clearance Center. Violations are liable to prosecution under the respective Copyright Law.
The use of general descriptive names, registered names, trademarks, service marks, etc. in this publication does not imply, even in the absence of a specific statement, that such names are exempt from the relevant protective laws and regulations and therefore free for general use.
While the advice and information in this book are believed to be true and accurate at the date of publication, neither the authors nor the editors nor the publisher can accept any legal responsibility for any errors or omissions that may be made. The publisher makes no warranty, express or implied, with respect to the material contained herein.

Printed on acid-free paper

Copernicus Books is a brand of Springer

Springer is part of Springer Science+Business Media (www.springer.com)

It is not the strongest of the species that survives, nor the most intelligent that survives. It is the one that is most adaptable to change.
Charles Darwin
70% of the Fortune 1,000 Companies will be replaced in a few years. Not because they didn't get enough fans on Facebook, but because they didn't adapt to the new networked society.
Brian Solis

Foreword

Dear readers,

Digital Darwinism—Certainly a pithy term to describe the changes looming ahead. But in the economy, it is about nothing else at the moment. Comparable to the **first industrial revolution** due to the invention of the steam engine in the middle of the eighteenth century and the **second industrial revolution** caused by the invention of electricity and the electrification involved towards the end of the nineteenth century, we are now in the middle of the **third industrial revolution**, driven by the omnipresent digitalization. This time, too—as in the previous revolutions—it's all about a profound and permanent re-creation of economic and social conditions. Our entire life circumstances and working conditions are changing massively.

For this reason, it is again all about **a fight for survival**, which companies successful for centuries have already lost, and the outcome of which for companies—previously—spoiled by success such as *Sony*, *Nokia,* and *Blackberry* still remains questionable. It thus becomes apparent that it is no longer about size, it is not necessarily about speed, and it is not only about the extent of conforming or strength. Today, it is more about:

> The survival of the smartest!

How can companies adapt to the rapidly and **radically changing market conditions**? How are the companies and their employees coping with the **rapidly changing business models**? For digitalization and social media are not primarily about changing company communication—as frequently presumed on the management level! Social media, big data, increasing digitalization, and much more are triggering an outright **tsunami**, which will destroy large parts of today's economy! This tsunami represents an attack on business models, sales concepts, marketing, communication, service, market research, as well as in general on the way we interact with customers and on markets! After this storm, nothing will be the same as it was before! We will have to **say farewell** to a lot of **practical knowledge** acquired with great effort!

This book was written to give food for thought about what **effects digitalization and social media** have on **established business models** and **successfully introduced brands**. It is our intention to waken up and at the same time set creative impulses so that new avenues can be explored, for the challenges lying ahead of us, our companies, and thus also directly ahead of our employees in the years to come are gigantic. The key challenge is to master the **digital transformation**.

So far, only a few managers were sanctioned for not being active in social media—in comparison to those who were reprimanded for the mistakes they made there. Yet that will change faster than many think. Then the **inactiveness in social media** will be reprimanded.

Erik Qualman says: "We don't have a choice whether we do social media, the question is how well we do it." The question about the **ROI of social media** is sometimes answered as follows: "The ROI of social media is that your business will still exist in 5 years!" This is why the term ROI in the context of social media is filled with additional content: ROI is to be understood as the **Risk of Ignorance**. Or as *Brian Solis* so aptly put it: "The End of Business as Usual" and "Engage or die!"

In any case, it is high time to first think and then act. And yet the core questions are: What effects will digital Darwinism have on our companies? And how does our **strategy for managing** the challenges lying ahead of us look?

- Ignore?
- Fight?
- Be overwhelmed?
- Or what else?

It is our aspiration to not only **raise exciting issues** but also to **convey concrete ideas** to **promote your creativity** and **stimulate solution processes in your companies**.

Once again, the saying goes: **Technology changes. Economic laws don't!**

So let us avail ourselves of the new technological possibilities to build up a **future-oriented success strategy** within the existing economic laws! The fact that more and more contents—generated by yourself and by customers—are available fully digitalized makes many technological restrictions become obsolete and creates entirely new fields of design. It is necessary to recognize these as early as possible to start the process of digital transformation.

One thing needs to be clearly defined at this point: tackling developments "revolving around social media" merely because everybody is is not enough. The main question for all companies even in this context is: Can we achieve **higher added value** for our own company if we deal with this issue? Marketing should make a decisive and value-adding contribution towards this—as should be adopted by every marketing concept. The Olympic principle—"It's the taking part that counts!"—thus does not apply here!

In the course of writing this book, we have realized the following five design ideas:

- **Fun**: Having fun reading—as it is, many of you will be reading (having to read) this book in your leisure time!
- **Food for thought**: Aspects to think about, to continuously stimulate your own considerations!
- **Think boxes**: Concrete questions to stimulate the process from becoming aware to acting!
- **Remember boxes**: Highlights worth remembering!
- **Quick wins**: Room for ideas that can be implemented straight away! This is where spontaneous ideas and solutions can be noted down right away so that they don't get lost.

Yet, with this book, we would also like to even convey **proposals for action and solutions** and inform you about **best practices**. And overall, we would like to encourage you to start with your **own finger exercises** as long as the market is not demanding ready-made concepts—and rather forgives the mistakes companies make! Yet, the way *Mark Zuckerberg*, founder and CEO of *Facebook* put it, still applies:

▶ "Done is better than perfect!"

Hopefully you will be able to say at the end of this book: yes, we feel fit to be able to survive the **era of digital Darwinism** or we now know ways of ensuring the necessary process of a **digital transformation**!

We wish you every success doing so!

Bonn/Königwinter, Berlin   Ralf T. Kreutzer
Cologne, NY   Karl-Heinz Land

Contents

1 Why the Digital Revolution Is Challenging You and Why You Have to Act Now...... 1
2 Digital Darwinism and the Social Revolution: What Basic Needs of Man Represent the Fuel of the Revolution on the Part of the Customer?...... 41
3 Big Data and Technology: Drivers of the Information Revolution on the Part of the Companies and Accelerators of the Era of Cooperation...... 77
4 How the Social Revolution Is to Be Managed...... 99
5 How Marketing Becomes the ROI Driver Within the Company... 129
6 Confidence: The Latest Currency in Marketing and Management...... 147
   Mom and Pop Store 3.0...... 151
7 Social CRM: The New Rules of the Game in Leading Customers... 163
8 Why Marketing Is Becoming a Service...... 193
9 The Necessity of Change Management: Why Our Traditional Communication and Organizational Structures Are Becoming Obsolete...... 209

Appendix...... 249

References...... 251

The Authors

Prof. Dr. Ralf T. Kreutzer is Professor of Marketing at the *Berlin School of Economics and Law* and a Marketing and Management Consultant. He worked in different leadership positions at *Bertelsmann, Volkswagen,* and *Deutsche Post* for 15 years before being appointed Professor of Marketing in 2005.

Prof. Kreutzer has set significant impulses concerning the various topics evolving in the fields of Marketing, Dialog Marketing, CRM/Customer Retention and Relationship Management, Database Marketing, Online Marketing, and Strategic and International Marketing by regularly providing publications and lectures and acting as consultant to a wide range of national and international companies.

His most recent publications include "Kundenclubs & More" (2004), "Marketing Excellence" (2007), "Die neue Macht des Marketing" (2008), "Praxisorientiertes Dialog-Marketing" (2009), "Digitaler Darwinismus—Der stille Angriff auf Ihr Geschäftsmodell und Ihre Marke" (2013), "Digital Darwinism—The silent attack on your business model and your brand—*the* ThinkBook" (2014), "Praxisorientiertes Online-Marketing" (2nd edition, 2014), and "Praxisorientiertes Marketing" (4th edition, 2013).

Contact:
Prof. Dr. Ralf. T. Kreutzer
Professor of Marketing at the Berlin School of Economics and Law and Marketing
 and Management Consultant
Alter Heeresweg 36
53639 Königswinter
Germany
Phone: +49 (0) 171–8668285
Email: kreutzer.r@t-online.de

Karl-Heinz Land is founder & CEO, Digital Darwinist & Evangelist at *neuland*. He is considered a visionary and advises companies concerning questions of digital transformation and vision.

Main foci: Social Media, Mobile, Big Data, Analytics, and e-Commerce. The repeatedly awarded entrepreneur and thought leader was accorded the title of "Technology Pioneer" by the *World Economic Forum* in *Davos* (WEF) and *Time Magazine* in 2006.

At *neuland*, his mission is to support organizations in managing the digital transformation successfully. *neuland* helps companies in making full use of the potential of the digitization. *neuland* assists in recognizing the need of change, adopting a digital strategy, and implementing successful projects of digital transformation.

Karl-Heinz Land has been working as Senior Executive, General Manager, CEO, and Executive Vice President EMEA in internationally leading Technology enterprises—incl. among others *Oracle, BusinessObjects, Microstrategy,* and *VoiceObjects*.

Contact:
neuland
STARTPLATZ
Im Mediapark 5
50678 Cologne
Germany
Phone: +49 (0) 221 97584041
Email: khl@neuland.me
http://www.neuland.me

Why the Digital Revolution Is Challenging You and Why You Have to Act Now

Until recently, we were able to laugh our heads off at the following definition of social media marketing as an important aspect of the digital revolution. **Social media marketing is like teenager sex**: everybody is talking about it, nobody really knows how it works, and when it happens, we ask: that was it?

Today, this explanation makes us choke on our laugh, because it no longer suffices to be satisfied with such explanations in the complex world of social media and make yourself s at home in the **circle of the ignorant**. For—according to the key results of a study by McKinsey (2012) interviewing 200 companies—the **relevance of social media** is considerable: After all, 70 % of the large- and medium-sized companies place high strategic significance on social media. But only 5 % are satisfied with their performance—in comparison to the potential social media offer as a whole (cf. McKinsey, 2012, p. 11).

Based on this study, there is a group of **social media pioneers**, which has been implementing a wide range of social media applications for an average of 2 years and, in doing so, determined an above-average impact on their business model (cf. Fig. 1.1). In contrast to that, there is the group of **social media newcomers**, who hardly use social media and thus also only reveal a limited background of experience.

In the acronym **SOCIAL,** *McKinsey* has clearly elaborated the differences between these two groups (cf. McKinsey, 2012, p. 12):

Strategy: The existence of a cross-corporate social media strategy based on strategic principles.
Organization: The engagement of full-time employees and provision of budgets for social media activities.
Criteria-based platform selection: Social media platforms were selected according to specific criteria—derived from the social media strategy.
Integration: The integration of social media activities in the entire value added chain.
Awareness: High awareness for social media in the entire organization.

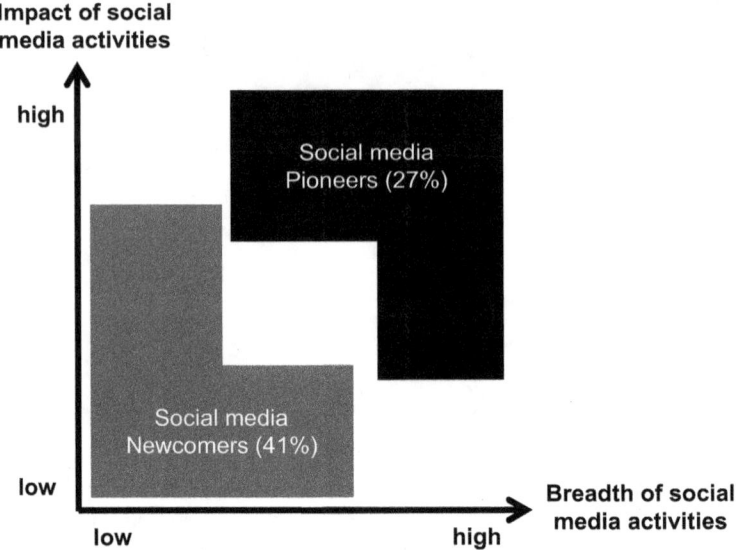

Fig. 1.1 Pioneers and newcomers to social media—McKinsey Social Media Excellence Survey ($n = 200$) (*Source* Author illustration, data source: McKinsey, 2012, p. 11)

Leadership: Involvement in social media with top management priority—high involvement of the senior executives.

Key question: Are the aspects mentioned already on the **agenda of the CMOs** (Chief Marketing Officers)? What are in their opinion the **central challenges** in the years to come? And how well prepared do the CMOs feel they are for these challenges? The answers to these questions are both enlightening and shocking at the same time! In the scope of the CMO study by *IBM*, 1,734 CMOs from 19 sectors in 64 countries were questioned about this. The **four greatest challenges** named by the CMOs were (cf. IBM, 2011a, p. 3):

- Data explosion.
- Social media.
- Growth of channel and device choices.
- Shifting consumer demographics.

These four central challenges alone underline the fact that existing **business models** and **established brands** can be **shaken to their very foundations** by the changes looming ahead. For this reason, it is important to ask how well prepared the CMOs feel for these challenges. Or to put it the other way round: how many CMOs do **not** feel sufficiently prepared? Figure 1.2 shows the figures. Based on our talks with many CMOs all over the world, we know that the situation has not changed significantly since the *IBM* study was published.

1 Why the Digital Revolution Is Challenging You and Why You Have to Act Now

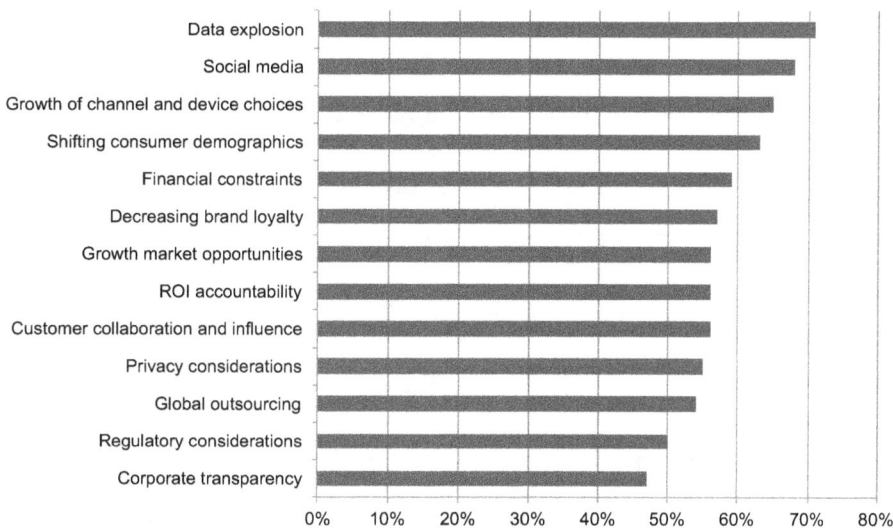

Fig. 1.2 Percent of CMOs reporting underpreparedness (*Source* Author illustration, data source: IBM, 2011a, p. 2)

In our view, the results show a dramatic call for action:

- 71 % of CMOs report underpreparedness according to **data explosion**
- For 68 % of interviewed CMOs, **social media** are a mystery
- The **growth of channel and device choices** is a **challenge for** 65 % and they have found no way to manage it yet
- And 63 % feel not well prepared for the **shifting consumer demographics**

What consequences these changes have in the classic marketing field of **communication** is shown here: according to research by *Procter & Gamble*, in 1965, three TV spots sufficed to reach 80 % of American women aged between 18 and 49. In 2012, as many as 97 spots were necessary to achieve this—and still counting (cf. Einicke, 2012, p. 9). Here we can see what direct consequences the **multiplication of media offers** as well as the **fragmentation of media exploitation** stimulated by this have.

But how can the necessary **change processes** be initiated in the companies when the top representatives of their kind—in this case the CMOs themselves—do not feel up to this issue? For especially in the case of the abovementioned **major challenges**, the CMOs themselves acknowledge having **major deficits**. In our opinion, this is an **honest** but also **daunting evaluation**. It calls for solutions, ideas, and information, in order to be able to stand on the winning side in the forthcoming or prolonged battle of choice. This book makes a significant contribution to survival in the era of digital Darwinism.

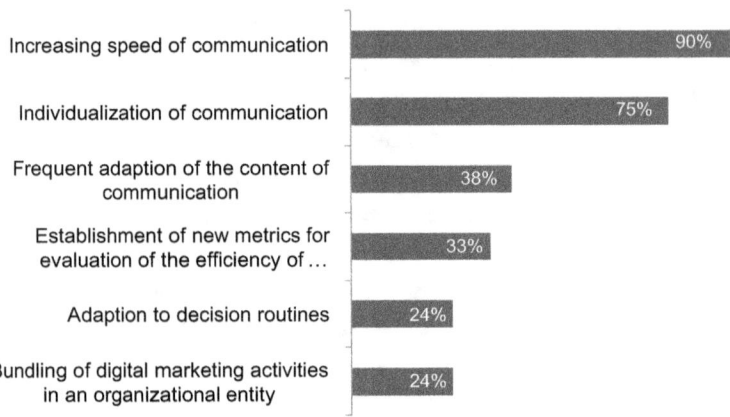

Fig. 1.3 Areas in marketing which are most affected by the Internet ($n = 100$ managers, multiple replies possible) (*Source* Author illustration, data source: Camelot Management Consultants, 2012, p. 19)

The results presented can be substantiated even further. A survey conducted in 2012 interviewing 100 managers in various sectors shows the prevailing concernment more impressively (cf. Fig. 1.3). For many respondents, the increased **speed of communication**, the **individualization** of it, as well as the **frequent adaption of the contents** are the most far-reaching changes. At the same time, it becomes obvious that there is a lack of **metrics**, in order to be able to capture successes and failures promptly. All of this, after all, imperatively leads to **changes in the operational and organizational structure of marketing** itself (cf. Chap. 9 for more detail concerning the digital transformation).

But how is the affinity to social media in our own company? How great is our **interest in the utilization of social media**—privately and for the company—differentiated according to company hierarchy? And what **power regarding the utilization of social media** do these different groups have for the own company? The **interest-power matrix** shows a distribution of the different key players typical for many companies (cf. Fig. 1.4).

It thus becomes obvious: Those employees with the greatest affinity to social media frequently (here, marketing employees) have the least power to promote the utilization of it for their own company. And the key players with the greatest plenitude of power (here, CMOs and CEOs) are usually the most reserved towards social media—a typical dilemma. Yet if a company is not attentive now, a **stairway to hell** is predestined.

> ▶ **Remember Box** The journey into the new media should not be started without the convincing commitment by the company management!

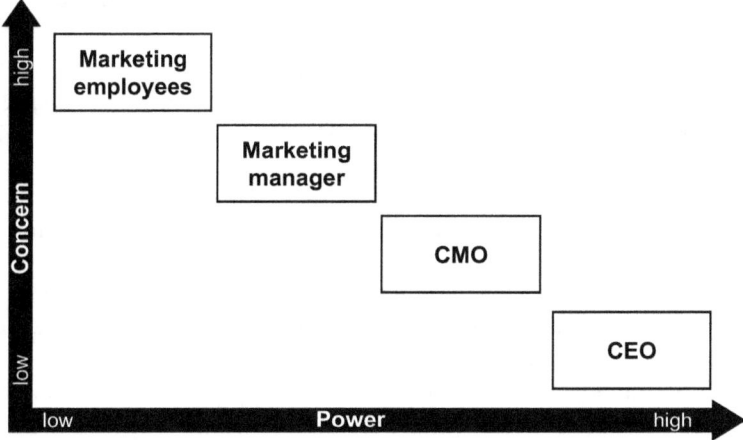

Fig. 1.4 Concern and power dimensions combined in a matrix—using the example of social media (*Source* Author illustration)

Think Box
- How are the social media dealt with in my company?
- How do I place myself in this matrix—and for what reasons?
- Who is putting the brakes on and who is encouraging the possible utilization of social media in my company?
- Which formal, substantial, and/or technical "defense arguments" are being brought forward in my company—and how sustainable are they?

So how great is the pressure to act for the own company now? An eye-opener for this purpose is the answer to the question How widespread is the **use of the Internet and social media** in today's households? Exciting results on this are:

- In 2012 about 274 M people in the USA use the Internet—in comparison to ca. 245 M in 2011 (cf. Statista, 2013a). According to Forrester studies, the **online penetration** rate of the US population is stagnating at 79 % in 2012 (as in 2011). **Daily Internet usage** is risen by 5 % according to a study with close to 60,000 US online adults (in 2011 78 % and in 2012 85 %; cf. Lardinois, 2012).
- Another study shows that in August 2011 78 % of US-American adults were online—in comparison to about 60 % in 2001 and 14 % in 1995. There are **huge growth rates in all age groups**. The highest increases are in the age cohort of the people older than 50 years. In June 2000 41 % of the 50- to 64-year-old US-Americans use the Internet, while at the end of 2012, 77 % does so. At this time also 54 % of the 65+ age group use the Internet—according to a study based on more than 2,000 adults (cf. Zickuhr & Smith, 2012, p. 4 f.; Pew Internet and American Life Project, 2013). The **segment of the silver surfers** is thus taking

on greater significance—the Internet has long passed being a domain for young people only.
- The mobile usage of the **Internet** increases as well. In 2011 more than 31 % of the US population uses the **Internet** mobile—in comparison to about 25 % in 2010—and it is expected that in 2015 almost the half of the population will use it in this way (cf. eMarketer, 2011). Another study—based on more than 2,000 interviewees—shows that already 55 % of all US cell phone owners—which are 88 % of the population—use the cell phone for going online (cf. Smith, 2012, p. 2).
- It is important that, with all these developments, the increasing use of **mobile terminal devices** such as tablet PCs and smart phones does not replace stationary access but facilitates new types of use. More than one third of US tablet owners and almost one quarter of US smart phone users use the tablet/smart phone as **second screen** while watching TV for looking up information (cf. Winslow, 2012).
- Another study with 2,540 Internet users (8–64), matched to the US Census with regard to age and gender, shows that 85 % use their tablet and/or mobile phone as **second screen while watching TV**. 66 % do so as **third screen** (cf. OPA, 2012, p. 26).
- In 2012 about 44 % of US-Americans had a **smart phone**—in comparison to 31 % 1 year before—which is mostly common in the age group of the 25–34 years old (cf. Google, 2012, p. 2, 6). In 2012, **tablet PCs** are used by 19 % of the US population—an increase by 100 % since 2011 based on a survey with more than 60,000 interviewees (cf. Lardinois, 2012).
- Thereby the behavior of the usage of a smart phone and a tablet is quite different. US-Americans use their **smart phone** most often for emailing (75 %) and browse the Internet for entertaining aspects (73 %), while they use **tablets** mostly for gaming (84 %) and searching for information (78 %; cf. Google, 2012, p. 16).
- The study by OPA (2012) shows that 94 % of **tablet users** weekly use it for accessing content/information, 67 % to going online, and 66 % for checking emails. In this study, gaming was named just on the fourth place with 61 %. Which screen is the more dominating one in each case is determined by the respective context.

This makes it clear that not only **user behavior** is changing, it also varies to an increasing extent depending on the respective **context**—and **change processes** are extending through all age groups. All in all, how can we recognize the associated **challenges**?

▶ In the beginning there was the word!

Which words challenge us today? New **buzzwords** are circulated every day—almost faster than we can comprehend and process their content. Some are forgotten fast. Others can be construed as **weak signals**, which indicate strategic changes at an early stage—and which we should pay particular attention to, because if we succeed in recognizing their relevance faster than others do and act accordingly, this can lead to a decisive **competitive advantage**. Yet which buzzwords should we attach particular importance to? **Disruptive society** is one such key buzzword. The fact behind it is that what applied for a long time won't apply much longer.

▶ More of the same!

Discontinuities and **structural interruptions** determine the corporate landscape. That is why one thing is for sure: the **digital revolution** is not imminent but is already unfolding the **power of creative disruption** in many sectors today. New stars are appearing in the corporate skies such as *Samsung, amazon, Google,* and *Facebook*, and stars that shone for a long time are disappearing. Who would have thought merely 3 years ago that *Nokia* would be pushed from the pedestal of the world's leading mobile phone manufacturer—and, at that, by a company that did not have any mobile phones whatsoever in its range until 2007, namely, by *Apple*? And who had *Apple* cut the wings off a few years before that with a product that previously hadn't been part of the computer manufacturer's service offering? *Sony*—cut down by an ingenious *iPod*! And who with their operating system for smart phones is driving *Nokia* and now even *Windows* in front of them? *Google* with *Android*—a provider that was not even active in this market until 2008!

If we analyze the causes of the success of *Apple*, then it is not only the convincing hardware that is the key to success. *Apple* in fact was successful thanks to *iTunes* as well as the setup of an extensive app store, setting up a so-called ecosystem. In this ecosystem, relevant offers are made to the customer in a closed system, which—from *Apple*'s point of view—is to be left as seldom as possible. This makes the following clear: by offering merely hardware, *Apple* wouldn't have temporarily become the world's most valuable company. It was the **consistent thinking in user cycles**—oriented towards **added value for the customer** that can be experienced immediately. As a consequence, competition can arise from diverse sectors both today and tomorrow—and frequently, only those who offer a (closed) user system with persuasive customer benefits in addition to attractive hardware have the lead. *Google* is creating an additional ecosystem—and it's always hard for us to go without!

Think Box
- Which sectors or from which companies do the greatest risks threatening my business model come from?
- Who has the potential—especially those who do not belong to my previous competitors—to break into my fields of services?

(continued)

- Where is my company vulnerable enough to alleviate an attack on my business model and my customer relations?
- Who has been entrusted with monitoring these developments—or who should be entrusted with this?

Figure 1.5 shows the **speed at which technologies are accepted** in general and also especially that of **social media** when gaining users. Whereas **radio** and **television** took 38 and 13 years, respectively, to gain over 50 million users, the **Internet** succeeded in doing this in 4 years, and the *iPod* in three. *Facebook* assembled a user community of 50 million after 1 year—and *Twitter* after as few as 9 months. The acceptance of *Google+* was even faster and was able to register 50 million users after only 3 months. And the driver behind this increasing speed of accepting technology? **A perceived relevance from the perspective of the users**.

This makes it clear that new technologies have never been able to gain acceptance as fast as they have now! The reason for the increasing **speed in accepting innovations** lies not only in the worldwide intercommunication which supports a diffusion of innovations across cultures, countries, and languages. The key condition for innovations to be accepted by wide masses is "**added convenience**"!

▶ Remember Box **Convenience** is the **driver of changes**. And convenience is also the **prerequisite for the acceptance of change**. In order to achieve this convenience for users, we can be guided by a guideline by *Steve Jobs* for *Apple*: **Simplify! Simplify! Simplify!**

For companies, **technologies** likewise represent opportunities if they rely on these technologies. However, technologies can also embody risks threatening companies, if the companies do not recognize their relevance for the users and do

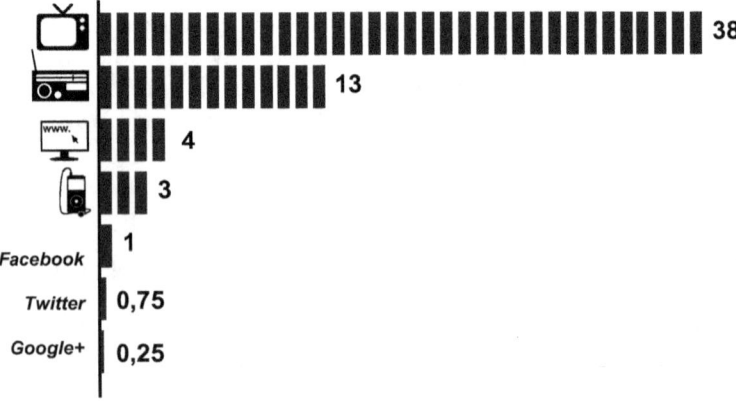

Fig. 1.5 How long does it take to gain 50 million users?—in years (*Source* Author illustration)

not rely on the corresponding technologies fast enough. Yet, which technologies should attention be directed to now—and which are to be disregarded? The annually updated **Hype Cycle for New Technologies** by Gartner (2013a) provides an important guide for companies. It shows which **phase of their life cycle** relevant cross-sector technologies are in. These technological life cycles are defined based on the **expectations** placed on the various technologies. This makes it visible which technologies are possibly overrated and which have already become an established tool or are turning into one (cf. Gartner, 2013a). In terms of the expectations of the technologies, *Gartner* defines five different phases which **shed light on the status of acceptance of new technologies on the market.**

Technology triggers
- In this phase, the first reports of the success of new technologies are published and readily taken up by the media. Whether these technologies will enjoy sustainable application cannot yet be foreseen at this early stage.

Peak of inflated expectations
- In this phase, success stories are published to further encourage the expectations of a new technology. At the same time, the first failures when using the technology become visible, which causes the expectations to reach their limits. The technological application still remains restricted to a few companies.

Trough of disillusionment
- This trough in the technological life cycle is based on the insight that many expectations of new "miracle weapons" were not fulfilled. In this phase, only those providers of technology survive who are able to convince the early adopters of the advantages of their technology in the long run. The other providers are ruled out of the competition.

Slope of enlightenment
- This is where it becomes increasingly visible how a technology can be utilized in a useful way. Technological developments of the second or third generation of the initial technology are offered and increasingly taken on by companies open for innovations and integrated in the workflow.

Plateau of productivity
- The technology is now used on a wide scale because its advantages are not only obvious but it also really pays off. The use as mainstream technology is preprogrammed. The use in more and more companies is now only a question of time.

In the Hype Cycle, *Gartner* also presents a **prognosis** as to when the plateau of productivity is likely to be reached. In the context of selecting the topics for this book, attention should be directed towards selected **technological developments**. The Hype Cycle shows that the topic of big data is on the brink of their peak of expectations. In the case of **big data,** it is assumed that it will have reached its plateau of productivity in approx. 5–10 years (cf. Chap. 3). The **Internet of things** will not only take a bit longer to reach its peak of expectations; it is expected that the plateau of productivity will not be reached until after 10 years.

Augmented reality and **NFC** (near field communication) are in the phase of consolidation as they have already passed their peak of expectations. **Virtual reality, gesture control,** and **predictive analytics** are already on their way to becoming a firm part of many corporate concepts and to accommodate the (reduced) demands made of them. In view of the dynamics prevailing here, every company should check for themselves what significance these developments have for their own company.

> **Think Box**
> - Which of the developments named in the Hype Cycle represent an opportunity for my company?
> - Where are risks to be expected, if any?
> - What further information is necessary in order to determine the possible effects of new technologies on my company?
> - Who is responsible for such an impact assessment of technologies?
> - With which measures can we react to particularly urgent challenges?

Gartner groups some technologies in so-called tipping point technologies, i.e., in technology groups that are at a watershed. These can have a huge impact on the business world or the entire society (cf. Gartner, 2012). This includes the so-called smarter things that are connected with each other via the **Internet of things**. But what exactly is meant by this "Internet of things"? This means clearly definable objects that are connected to each other via the Internet. Newer types of communication such as via RFID (radio frequency identification) or now via NFC (near field communication), i.e., wireless communication, alleviate the exchange of information and, in the case of products, can replace the use of bar codes. If objects are equipped with radio tags, then the data received in this case can, for example, determine whether an offer in a store is running out—and where necessary, an order can be automatically triggered. If in everyday life, people are equipped with these radio tags (e.g., via their smart phone), they can easily be identified and localized and thus highly individualized messages are sent to them—based on known preferences.

Ashton (2009), the inventor of the term "**Internet of things,**" has summarized the key contents precisely: "Today computers—and, therefore, the Internet—are almost wholly dependent on human beings for information. Nearly all of the roughly 50 petabytes (a petabyte is 1,024 terabytes) of data available on the Internet were first captured and created by human beings—by typing, pressing a record button, taking a digital picture or scanning a bar code. Conventional diagrams of the Internet include servers and routers and so on, but they leave out the most numerous and important routers of all: people. The problem is, people have limited time, attention and accuracy—all of which means they are not very good at capturing data about things in the real world.

And that's a big deal. We're physical, and so is our environment. Our economy, society and survival aren't based on ideas or information—they're based on things. You can't eat bits, burn them to stay warm or put them in your gas tank. Ideas and information are important, but things matter much more. Yet today's information technology is so dependent on data originated by people that our computers know more about ideas than things.

If we had computers that knew everything there was to know about things—using data they gathered without any help from us—we would be able to track and count everything, and greatly reduce waste, loss and cost. We would know when things needed replacing, repairing or recalling, and whether they were fresh or past their best. We need to empower computers with their own means of gathering information, so they can see, hear and smell the world for themselves, in all its random glory. RFID and sensor technology enable computers to observe, identify and understand the world—without the limitations of human-entered data. . . . The Internet of Things has the potential to change the world, just as the Internet did. Maybe even more so."

The multitude of information that can be generated via the Internet of things will continue to reinforce the trend towards **big data**. Data from various sources, generated by mobile or fixed-line access, are being increasingly interlocked via uniform protocols (in particular the Internet protocol IP). This makes data of a previously unknown quantity and quality available for analyses. The combination of such data streams with intelligent analysis tools—applied in real time—enables highly personalized customer approaches. Essentially, it is about the presentation of specific offers that not only match **a user's profile**. A good CRM—i.e., a customer relationship management—has been able to do that up till now. The opportunity being offered here rather entails directly matching the contents to be communicated with the respective **context of the user**—in time, space, and content (cf. Chap. 7). How important timing and context can be is illustrated by the following example: The information that there is a speed camera in Lombard Street in San Francisco is not very useful if this information does not reach me until my speeding has already been recorded on a police photo. Yet, if the information reaches me a few minutes earlier because the analysis systems have recognized that I am on my way into Lombard Street, the relevance of this information increases dramatically. However, this only applies if I am not traveling by bus, taxi, or bicycle. On the other hand, should I be traveling around New York, such information is of no relevance to me whatsoever.

> **Food for Thought** The value of information changes with time and space—and consequently with the respective context.

Many business models are still based on **static information structures**. To this end, information is gathered, stored in databases, and updated—annually or never (as experienced in major customer projects). Today, however, we have the opportunity that remains untaken to distinguish oneself in competition by accessing **dynamic information structures**, because offers can be presented with an unachievable one-to-one precision. For now, there is not only more and more information available about potential customers and customers but also more and more precise information. And this is happening at a speed that, until recently, was deemed impossible, facilitating **real-time marketing** (informational). At this point, may we refer to *Facebook*, the world's largest and best (because daily) maintained database of preferences (cf. Chap. 7 for more details).

The question we should be asking ourselves is: How can we make use of this **information to create added value for customers**, so that the customer generates greater **added value for our company**? In Japan, streams of passers-by are already being screened in order to adapt the contents of the large screens depending on the insights gained by this (cf. Chui, Löffler, & Roberts, 2010, p. 1). This, in turn, can accomplish something: a **greater relevance of the contents of the adverts broadcasted**.

Hence, what is the new challenge?

In Search of Relevance!

If the preferences of customers can be analyzed in real time combined with a certain location where the person happens to be, **dynamic triggers** (e.g., place of purchasing, price, and product availability) can be sent by mobile, in order to send the all-deciding trigger to purchase—at exactly the right moment and at the right place. The basis for this is, of course, that we have received **permission** from our customers to contact them in this way. **Check-in services** such as *Foursquare*, where you check in at certain locations (e.g., at *Starbucks*, at *McDonald's*, or at the airport), actively make such information available to the users.

> **Remember Box** Our customers and potential customers are permanently leaving behind **digital footprints** on all channels: by mobile, in stationary shops, when surfing on the Internet, and during activities in social media.

Even if, in the course of neuro-marketing, the long sought-for **buy button** has still not been found in the head of the customer, one thing does hold true: by creating a proximity of promotional influence in time, space, and content, the **relevance of our message** for the recipient increases dramatically—and with it, the probability of purchasing. This correlation becomes clear in Fig. 1.6. The great challenge consists in reading the **single point of truth** from our target person's

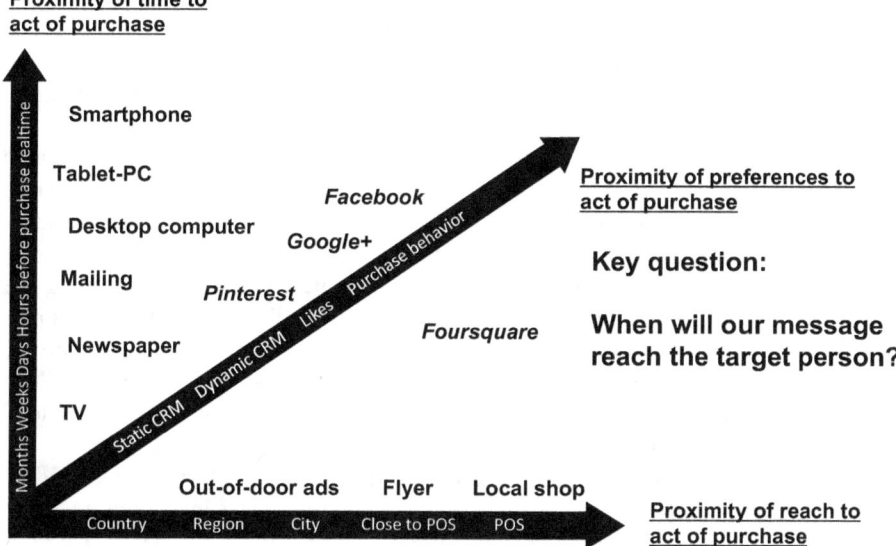

Fig. 1.6 Relevance of information based on its proximity of time, preference, and reach to the target person—360°-CRM (*Source* Author illustration)

digital footprints, in order to know what really counts for that person in the respective context.

Figure 1.6 shows that the **relevance of one piece of information** increases with the proximity in time, space, and content of a message. The **proximity of reach to the act of purchase** can be increased by playing out advertising attuned to the respective location of the target person (**location**). The permission to locate the user via GPS or by the check-in services previously mentioned makes it possible to determine the user's location more and more precisely. The **proximity of time to the act of purchase** is still frequently restricted with TV and radio advertising but also with newspaper ads and direct mail. A stationary, but in particular a mobile online presence, can reveal a much greater proximity in time to an offline purchase (**timing**). Finally, vital importance is also attached to the **proximity of preferences to the act of purchase** (**preferences**). Together, all three dimensions map the respective context in which the target person is in.

In static CRM systems, the data on the customer was only updated at larger intervals or on the basis of customer surveys. Dynamic CRM systems, in contrast, aim at continuously collecting the customer activities and to take them into direct consideration when addressing them. In particular with *Facebook*, *Google+*, and *Pinterest*, there is very up-to-date preference data, which is documented via "Likes," "+" and "Pin it." In accompaniment, information about purchases made is also being increasingly provided. Companies that succeed in pooling these three "proximity-generating poles" of location, timing, and preferences will always be ahead in the communicative approach. A **three-dimensional CRM** becomes possible. A decisive factor for this is that we have created a **single point of information** (also called single point of truth) in the company where the various streams of information converge.

The successes that *Tesco* has achieved with their customer retention system by playing out personalized coupons at the POS show the potential of such an approach (cf. Gallagher & Zoratti, 2012, p. 171 f.). However, the concept demonstrated above goes yet a decisive step further. Whereas the provision of coupons is based on an extensive analysis of the previous purchasing behavior of *Tesco* customers and the coupons are handed out at the POS, the concept revealed in Fig. 1.6 goes much further: not only the acts of purchase and preferences in the own shop are considered but also those, for example, that become visible in social media or by check-ins at certain providers. The corresponding permission granted by the user is always a prerequisite. Apart from that, the delivery now takes place via mobile—and ideally reaches the recipient at the point in time, at the location, and in the mood where the **greatest receptiveness for a communicative trigger** is given.

This **informational triad** is embodied by the term **three-dimensional CRM**, which simultaneously merges timing, location, and preferences in a communicative manner. This triad can be best made use of with **location-based services** (**LBS**) by linking the extensive database with a convincing, creative idea that offers the customer real added value.

> **Think Box**
> - How close are we to our target persons in informational terms?
> - How well do we succeed in collecting the preferences of the potential customers and customers—with high topical significance?
> - Can we determine the location of our potential customers and customers?
> - How can we use this knowledge about the "location" to increase our sales—even if we are possibly a mere online provider?
> - How close "in timing" are we to purchase decisions of our customers?
> - How could we achieve this proximity of time?
> - Put your company or your offers in Fig. 1.6. Add yourself to your relevant competitors now!

A further driver for serious changes shall be mentioned here: **mobile communication**. Figure 1.7 shows how mobile devices have developed. The development of mobile phones weighing several kilos to smart phones has opened up completely **new fields of application and use** for companies and customers. And this process will continue to increase because more smart phones are sold in many countries than standard mobile phones. This will cause another trend to increase dramatically: the increasing intensity of **using various platforms by mobile**. Not only is *Facebook* already used by more than 60 % by mobile, even online purchases and online research are being conducted to an increasing extent by mobile. With *Google* spectacles presenting important information directly and instantly to our field of

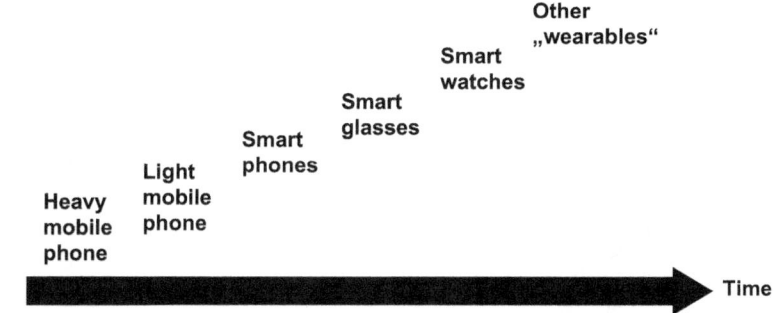

Fig. 1.7 Development of the "mobile terminals" (*Source* Author illustration)

vision, the next generation of mobile devices is in the offing. In addition smart watches try to attract our attention.

Studies shows what significance is to be attached to **mobile data traffic** today and in the future. A 12-fold increase in data transfer from mobile applications is expected by 2017—when taking the year 2012 as the starting point. The global mobile data traffic per month (in exabytes) will increase from 0.9 EB in 2012 to 11.2 EB in 2017 (cf. Cisco, 2013, p. 5).

The increasing prevalence of high-performance mobile terminals will inspire a further trend—the development towards a **cashless world**. It is already becoming apparent today that the entire payment infrastructure will change dramatically in the years to come. Taking a glance at the previously mentioned *Gartner's Hype Circle* shows that NFC (near field communication) payment has only just started off on its way there. In addition to the payment function, further functions are becoming available on the smart phone via apps (cf. Fig. 1.8).

One thing can be determined at this stage: we are at the beginning of a development that is allowing the smart phone to become a **smart service terminal** that will become a key holistic **control and navigation instrument** in the online and offline world. Car keys, passport, credit cards, coupons, purse, and much more will become obsolete, thanks to the smart phone—as normal as mobile shopping and mobile banking has already become for many (cf. Fig. 1.9). This will lead to purchasing behavior continuing to change dramatically. This process will be additionally reinforced by the fact that more and more information of relevance to purchasing processes is becoming available by mobile. The easy access to the multitude of information on the Internet takes place by **reading out bar codes**, the use of apps to **recognize physical products** (e.g., via the *RedLaser* app), the **identification of music** (by the likes of *Shazam*), as well as the multifaceted **application of QR codes** (such as with QR shopping with *eBay*). QR codes are already being used by *PayPal* in the payment terminals in the stationary retail trade. They are scanned and read out via the *PayPal a*pp in order to initiate an immediate payment transaction.

It appears that conventional applications are now turning into apps for our smart phones. Thus the development into an **app economy** is predetermined. A lot will evolve from the analogous world into the digital world. The credit card, for

Fig. 1.8 Physical objects will be substituted and digital ones used (*Source* Neuland, 2014)

Fig. 1.9 Variety of performance of the smart service terminals (*Source* Neuland, 2014)

example, today being a piece of plastic, will become an app. Due to this, it loses its physical limits. This means that in the future, a user can enable a third party to access this "software credit card" without having to actually let go of it. The development of the mobile phone to a smart service terminal is reinforced even more by the **trend towards a digitalization across added value**. Not only **data** (e.g., about our customers) and **processes** (such as consulting, sales, and payment processes) are being increasingly digitalized and thus made available via mobile but also the predominantly **physical products** provided up till now (e.g., books, newspapers, magazines, CDs, and DVDs) are losing their physicalness. This means **physical limits** are being overcome, which had been of great significance to our business models up till now and frequently represented the basis of them. Where books, newspapers, magazines, CDs, and DVDs had to somehow be **transported physically to the customer** in the past, this can be done away with in many cases today.

This development is described by the terms **zero gravity thinking** and **dematerialization**. Objects lose the physical restrictions they had in the real world when these objects are transmitted into cyberspace. The music and newspaper business demonstrate what long-term effects this zero gravity thinking has on entire industries. With this in mind, every company should delve into how not only their own communication but also the products and/or services they offer can be made available via mobile.

But even the **digital delivery to the buyer**, in order to physically store contents on the terminal, is no longer required to the extent that the contents are stored in the **cloud** and not made available until the first moment of use by means of **streaming**. Decentralized data storage per user is thus replaced by centralized data storage in the cloud. This trend towards relocating in the cloud is not only restricted to additional processes and entire business applications; digitalization and therefore the dematerialization encompass complete business processes.

> **Think Box**
> - What impacts will it have on my company and the relations to my stakeholders if smart phones do actually become smart service terminals?
> - How can we proceed to at least prepare ourselves for this development with "finger exercises"?
> - To what extent is our communication already accessible by mobile?
> - Which products and/or services can be made available by mobile?
> - What impacts does the extensive digitalization have on our customers or on their customers?
> - What repercussions are associated with the provision of our services and thus with our value added chain?
> - What impacts does this in turn have on our buyer behavior?
> - Where can we, as customers, benefit from this increasing trend towards digitalization?
> - In which areas can we generate added values by digitalizing contents?
> - Of what relevance is the increasing advance of cloud solutions to my company?

(continued)

- Where can we benefit? Where is more of an attack on our business model recognizable?
- Who could take on responsibility for that in my company?
- Has the time come for us to appoint a chief digital officer?

The entirety of these developments leads to a **new definition of business portfolios** across sectors. **Online retailers** are becoming **hardware manufacturers**, in order to cover a larger share of the (digital) value chain by means of setting up their own ecosystem. Thus *amazon* offers the *kindle* e-book reader at less than cost price to earn money from the sales of digital products. Other **online service providers** are turning into **software providers** (here *Google* with *Android*) and also partly into **hardware providers** (e.g., *Google* with *Nexus* in the smart phone and tablet PC market). At the same time, **hardware manufacturers** are becoming **portal providers**, as is the case with *Apple* with *iTunes*. Finally, **network operators** are becoming **content providers**, as can be observed with several telecommunication service providers. *eBay*—previously a **digital pure player**—is testing the pop-up store (a store that is only open for a few weeks) under the term *salesroom*. Here, real products are presented by different and of course well-rated sellers from the *eBay* platform in a real store. The products presented here can be bought directly via the *PayPal* app by scanning the QR codes containing information about the price and availability. Payment is effected via *PayPal*, and the goods are sent to the desired delivery address. The keyword here is **QR shopping** (cf. PayPal, 2012).

In addition, service providers are opening up completely new service sectors as is the case with the cloud services at *amazon* and *Google*. And **artists**, whether they are musicians, graphic designers, authors, or amateur craftsmen, are becoming **direct marketers** of their own creations. They no longer need the classic sales structures in many cases (keyword: "self-publishing") or are using completely new **online sales platforms** in order to market their creative fruits. These include, for example, the world's largest marketplace for "any kind of hand-made products" *etsy.com*. This marketplace has more than 30 million members, and in 2012, sold fashion, furnishings, and all kinds of paraphernalia worth around 900 million US $—and counting rapidly (cf. Etsy, 2014).

The concept expressed by these online sales platforms is called **long tail** today, which was coined significantly by Andersen (2009). Figure 1.10 makes the origin of the term "long tail" comprehensible. In order to draw this curve, it is necessary to enter the objects of investigation (e.g., products, service) on the x-axis placed in reverse order according to the number of sales made. So, the books by *E. L. James, Stephenie Meyer, and Nicholas Sparks,* for example, which top the bestseller lists in the USA as blockbusters, can be found leftmost on the x-axis. The perfect selling German version of this book, for example, can be found at the beginning of the long tail and works about the "lovelife of bees in the Middle Ages" possibly only find a number of buyers of between 0 and 10 and are located at the end of the long tail. The

1 Why the Digital Revolution Is Challenging You and Why You Have to Act Now   19

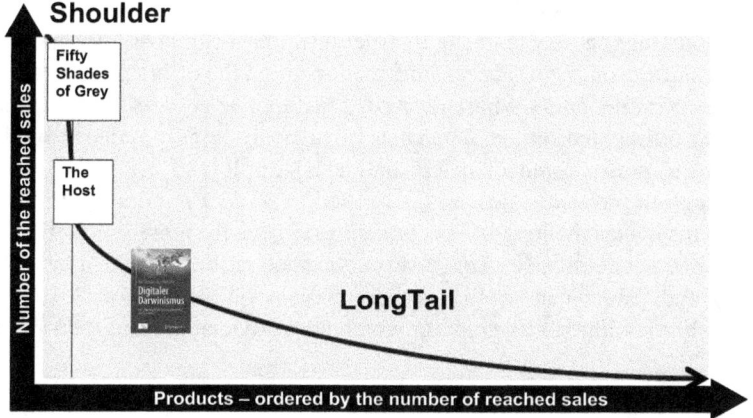

Fig. 1.10 Long tail concept—example of the market for books (*Source* Author illustration)

respective number of sales achieved (as number of value) is entered into the y-axis. The first part of the curve is called the **shoulder**: here are the bestsellers or the objects of investigation also called blockbusters. Besides products (such as books, clothing, music, or films), this can also be services. The second part of the curve is called the **long tail**. Here you can find all other offers enjoying significantly lower demand. Here one can aptly speak of the **mass of niches**.

The fact that today highly specialized providers can be run by looking after the smallest of segments and market niches is owed to the Internet. The Internet, and in particular the functions of search engines, can bring together providers and demanders of **niche products** at reasonable costs. Whereas totally unusual music, books, or clothing demands cannot be satisfied economically in many cases by the classical stationary sales channels with their limited regional catchment area as well as their limited range of products, the Internet opens up lucrative business fields. In extreme cases, there is a global demand for highly personalized offers. Overcoming regional borders leads to the **development of lucrative niche markets** and thus to the long tail (cf. Fig. 1.10). Essentially, three **modes of action of the long tails** can be elaborated according to Andersen (2009, p. 60–67), which have led to a reduction in costs, in order to serve niche markets in a profitable way:

- **Democratization of resources**

    The extensive distribution of important production means (e.g., PCs, MP3 recorders, digital cameras, and mobile camera phones) as well as do-it-yourself products of all kinds makes it possible nowadays for millions of people to make their own creations. This means that many new texts, pieces of music, photos, and videos as well as other products, which are available for sale, emerge every day. This elongates the curve in Fig. 1.10 to the right, i.e., the long tail grows.

- **Democratization of sales**

  Anybody having access to the Internet nowadays can post information about their own offers online or access information about the offers presented on it and order them immediately where desired. This applies equally to products available both offline and online. The costs of sales are rapidly reduced by this fact because the presentation of offers online is simple and affordable or free of charge and no physical sales area is needed to serve a regionally limited target group. This makes the long tail become thicker because more transactions can be carried out economically. The drivers for such offers are the likes of *eBay*, *iTunes*, and *amazon* as well as other relevant online platforms such as the previously mentioned *etsy.com*, on which anybody can present their products.

- **Connection of supply and demand**

  The Internet alleviates the pooling of supply and demand particularly by means of search engine and social networks as well as spanning through social bookmarking, blogs, forums, and communities, with which personal recommendations are voiced. By means of the information located and communicated here, it is becoming increasingly easier to bring together niche providers and searchers for niche products. This means the demand for mass products can be shifted from the shoulder area to the area of the long tail because, as an alternative to standard products, offers that may possibly fulfill perfectly one's own requirements become detectable.

Even if the long tail approach describes the relevance and profitability of **serving niche markets** in a comprehensible way, it does not render the **Pareto principle** (also called the **80:20 rule**) invalid. Figuratively, this expresses that there are **concentration effects** in all fields, therefore also when purchasing products or with the demand for services. It may indeed be the case that the total sales in niche markets exceed that of blockbusters, yet it must be pointed out that behind a blockbuster such as *Harry Potter* there is exactly one author and one publisher, whereas there are a number of providers behind the offers of the long tail; it is thus hardly constructive to summarize their turnover without summarizing the marketing costs across all providers. Furthermore, there will always be **the-winner-takes-it-all models**, which due to their size dominate many other providers and thus outrival them.

Many examples of such concepts, which with a high concentration of users involve one or a few providers and thus reveal the sketched **Pareto effect**, can be found particularly on the Internet. Such concepts can be found not only in search engines where it is interesting for most companies to be present on the largest provider (in many countries *Google*). In social networks, niche products will also have little success because people feel attracted to the networks where many of their friends already are. **The-winner-takes-it-all effect** appears this way on, for example, *Facebook*. And even *Google* is finding it difficult to successfully position its social network *Google+* against *Facebook*. Even powerful providers—such as *Microsoft*, which was not able to score in either social networks or the search engine sector—find it difficult to fight against such effects. Even the named provider *etsy.com* represents the-winner-takes-it-all concepts with respect to their sales function. For if I, as a provider, wish to present myself to many potential customers, then I should go to where there are many demanders—at *Etsy*, for example, more than 60 million visitors per month (cf. Etsy, 2014).

Nevertheless, it is important to point out that across the sales bandwidth of the Internet, a so-called trickle-up effect can come to bear. Until now, only a **trickle-down effect** was spoken of, whereby a **trickle-down effect** "from top to bottom" is meant. Originally this primarily referred to the development of wealth in countries, whereby this wealth—ideally—trickled down from the rich to the classes of society below little by little (as currently in China and India). Nowadays, this term is also used to show how, for example, a new strategic focus in a company or the commitment to "green issues" in the sense of a stronger ecological focus of a company is implemented in the entire company little by little. A **trickle-up effect**—against gravity as it were—can be spoken of in the case of selected offers on the Internet because offers presented in the niche can manage to become visible to the global public by means of the diverse communication tools of the Internet. These can be songs or texts of previously unknown artists, whose popularity can be significantly increased by viral effects on the Internet within a short period, as was the case with *Justin Bieber*. Even the *Fifty Shades Trilogy* was first published online by the author. It thus becomes apparent that the long tail permeates to the shoulder. And many business processes that were created from the long tail many years ago have now reached the head. In this context, we must think of the likes of *amazon*, *eBay*, and *YouTube*!

The drivers behind these developments are not only the providers of new technologies but also the **altered behavior of the customers**, partially as a consequence of them. However, these changes are no longer limited to the **digital natives**, even if their behavior patterns in particular have shifted significantly. Digital natives gain their first Internet experience "before birth"—based on the behavior of their parents—are in possession of their first mobile phone at the age of five, operate their parents' *iPad* at the age of six, and set off on their first extensive excursions on the Internet at the age of seven, hopefully under the watchful eyes of their parents. And if toddlers do actually pick Mom or Dad's newspaper, they may mistake the paper bundle for a "broken *iPad*"—because it doesn't function like this!

What impacts do these developments have on the groups of younger customers? All in all, the majority of youths up to the age of 20 have already spent thousands of hours playing computer games and, in doing so, adopted specific ways of thinking and behaving, which are totally alien to the older generations. The **reservoir of experience** of the various generations is thus filled with completely different contents. What impacts is *Moshe Rappoport*, Executive Technology Briefer of *IBM*, expecting in the light of this development? Analog to the behavior with computer games where one gets to the end faster, thanks to risky behavior, and after "game over" simply starts from the beginning again, the young generation is characterized by an **increased readiness to assume a risk** and **act fast**. At the same time, a **"reboot mentality"** is emerging. The users can drop out of this "reboot world" at any time—loyalty is a currency of days gone by, at least if it was generated with instruments of days gone by or rather enforced by these. This is being tried by dozens of providers of plastic cards and other customer retention programs, some of which no user can or wants to figure out, because it would be necessary to go through and memorize an 80-page manual! According to *Rappoport*, this also goes for Internet platforms, for example: "You go there and look to see if it is funny; you discuss, chat, and lose interest and go on to the next project. Someone else comes along with a different message, with a trick, with other things. With young people, the need for steadfastness is not so pronounced. 'Come and go', 'Press the red button and restart' characterize virtual life" (Sohn, 2012).

What will happen now if this virtually implemented "**come and go mentality**" is also implemented in everyday reality? Even if the previous generation coined the term "life partner," is the new term then "temporary life partner"—or somewhat catchy—"hourly companion"? The behavior pattern acquired online also plays a key role in the acceptance of new products and technologies: **Try what you like!** And if you're not satisfied, the relevant provider is mercilessly knocked off their pedestal. And it can hit anybody for that matter, as the examples of *Nokia*, *Blackberry*, and *Sony* show.

> **Think Box**
> - What possibilities does our business model offer to "weave in" the customer in an extensive ecosystem?
> - How can our added value chain be further developed in the light of these challenges?
> - What consequences does the long tail concept have for us? Does it offer opportunities or rather risks?
> - Can we continue to benefit from the Pareto principle—or can we lever it out by means of the "power of the niches"?
> - How can we succeed in benefiting from the changing behavior patterns of the digital natives?
> - Where in my company can the processing of these questions be outsourced to?

An intensification and additional dynamics have experienced all of the previously discussed challenges by means of a phenomenon described by the term **social media**. This term classifies online media and technologies, which enable the Internet users to conduct an exchange of information online that goes far beyond classic email communication (cf. Kreutzer, 2014, p. 378). This means that, for the first time ever, all classes of population and all stakeholders that a company deals with are holding an **extremely powerful** and **extremely high-publicity communication tool** in their hands (cf. Fig. 1.11). At this point, the indication that social media can reveal **value-creating** and **value-destructive** contents is important—and it is up to the company's own involvement as to which contents dominate!

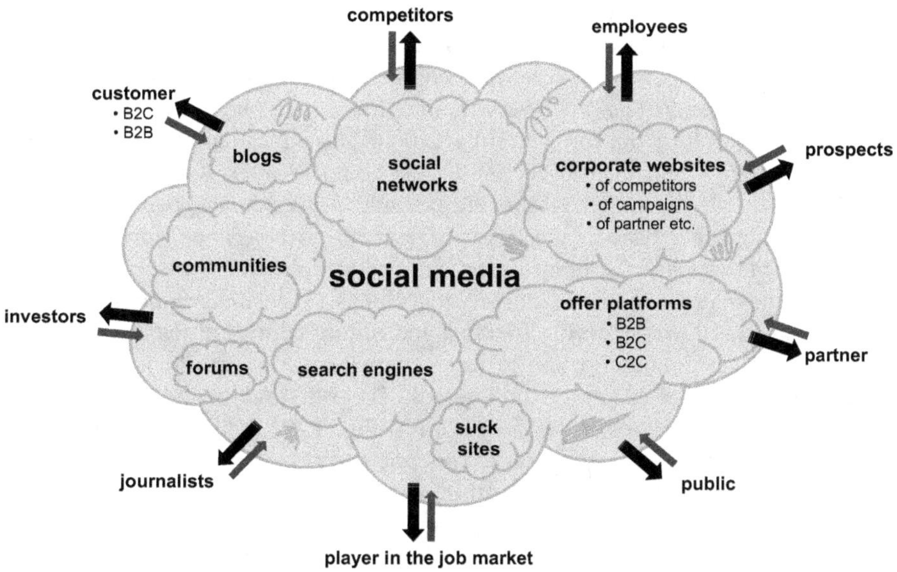

1 Why the Digital Revolution Is Challenging You and Why You Have to Act Now

Fig. 1.11 Increasing uncontrollable and complex building of opinions of stakeholders through social media (*Source* Author illustration)

The **relevance of social media has changed**—which can be measured by the development of appropriate queries on *Google*. Here, a **search dynamic** becomes visible that until now, only a few phenomena were blessed with. The volume of searches across the world has increased tenfold since the middle of 2009. The keywords **social media marketing** and **social marketing** dominate. **Social media** include in the first instance **social networks** such as *Facebook*, *LinkedIn*, *Pinterest*, and *Google+*. Even their development involved structural breaks. Who would have guessed when *Facebook* started up that this company would succeed in bringing more than one billion users to take part in as early as 2012 and thus win over almost half of all Internet users worldwide? Who would have dared to have forecast that users worldwide, beyond educational, gender, age, and cultural limits, would once be ready to make public a **multitude of data** about themselves so **up-to-date and in such detail** that was previously unimaginable—which can often be viewed by the whole world. This online presence of the users is frequently supported by a multitude of text, music, as well as still and moving image files, which enrich their own profiles. At the same time, an **update speed** of the contents presented there becomes visible, which is often not only on a daily basis but rather on an hourly basis. For many centuries, marketers used to dream of such a vast amount of data—nowadays described by the previously mentioned buzzword **big data**—and data protectors have had "nightmares" about it.

Another part of social media are **media sharing platforms**, the most important of which being *YouTube*. *YouTube* itself does not create contents but merely makes the platform available for interexchange and is financed, among others, by promotional offers labeled as "sponsored videos." The community of the *YouTube* users creates and controls the contents on *YouTube* and decides what is popular. Over six billion hours of video are watched every month on *YouTube* and 100 h of video are uploaded to *YouTube* every minute. More than one billion unique users visit *YouTube* each month (cf. YouTube, 2014). These facts and figures cover the contents created both by the users and by the companies. Of what relevance this channel is becomes clear in the following example: *Felix Baumgartner's* jump from the stratosphere in 2012 was the largest live event on *YouTube*. More than eight million people watched the jump on the Internet—in real time!

Further developments of social media are **blogs** (like *Tumblr*), **micro-blogger services** such as *Twitter*, **online forums,** and **online communities**. Social media have changed the communication between companies and customers massively—especially among the customers themselves. The **1st stage** of the differentiation between communication and performance (cf. Fig. 1.12) dominated well into the 1960s and 1970s of the past century. This **one-to-mass orientation** comprised a widely **undifferentiated communication**, for example, in the mail-order business by means of only one main catalog per season as communication with the customers. At the same time, the **marketing of standardized offers** dominated. This company behavioral pattern was supplemented or replaced in the **2nd stage** in many areas by the **one-to-many** principle. The **target group-specific communication** accompanying the possibilities of dialog marketing was characterized by terms such as customer base marketing, database marketing, or relationship marketing. More flexible production structures enabled at the same time—at least to a certain extent—a differentiation of performance based on customer expectations.

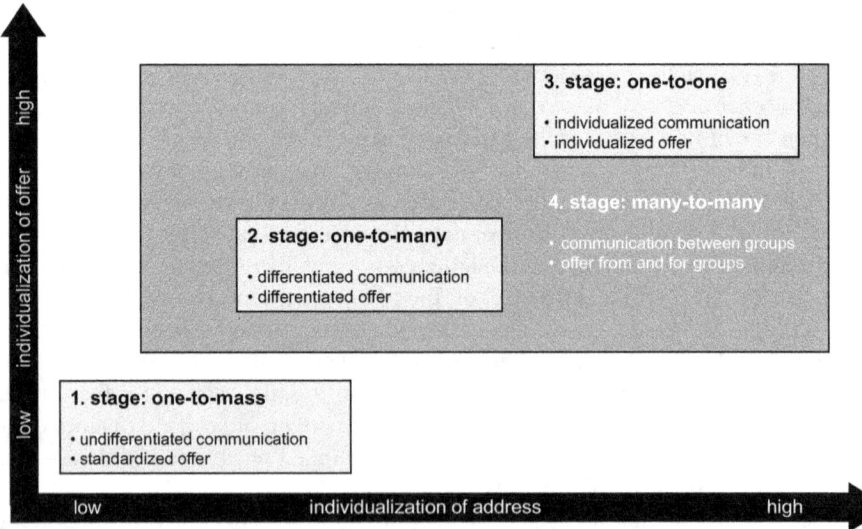

Fig. 1.12 Change in degree of the differentiation of communication and performance (*Source* Author illustration)

In the 1990s, under the key term **customer relationship management** (CRM), the step towards the **3rd stage** of **one-to-one** was systematically prepared in some areas—based on more sophisticated database, higher performance analysis systems, and further developed technology in communication. In doing so, it was attempted to increasingly put potential customers and customers as individuals in the center of the communication and to personally address and support them. The personalization and individualization of addressing them is based on the specific knowledge about the person and/or information about the history of the relationship between the person and the company. An excellent example of how this type of communication is put into perfection is shown by customer retention programs like *Payback* in Germany: approx. eight million different printed versions are being produced for distribution campaigns of (paper-based) coupons with a circulation of nine to ten million copies (cf. Schmidt, 2012). However, only a share of the companies is exploiting the possibilities that professional dialog marketing offers the companies (for more details cf. Kreutzer, 2009).

With regard to **corporate communication**, it is necessary to define the corresponding degree of individualization depending on the respective business model. In the case of a national provider, all three types of communication could be used:

- **One-to-mass addressing** (e.g., national TV campaigns and/or advertising in widely circulated magazines and newspapers)
- **One-to-many measures** (such as adverts in special interest magazines as well as mailing selected target groups on the basis of rented addresses)
- **One-to-one addressing** (e.g., based on the specific history between potential customers or customers and the own company)

Each form of communication therefore has its eligibility. The approach is to be derived from the respective communication goals to be reached.

Frequently, major production-technical boundaries are set on the **individualization of performance**. In the case of specific business models, a one-to-one orientation is successful, such as classically at the hairdressers or at furniture manufacturers. But also providers such as *Dell* as a supplier of customized and configured computers, *scenterprises.com* as provider of personalized perfumes, and *cerealize.com* as source of supply of custom mixed cereal are successful in this. In contrast, highly praised innovations such as personalized running shoes with a personalized logo on them turn out to be standard shoes after taking a close look. For this reason, it is imperative to differentiate between the **individualization of addressing** and the **individualization of performance** when it comes to **one-to-one marketing**.

Nowadays, we see ourselves being confronted with the **4th stage** of communication: **many-to-many** (cf. Fig. 1.12). This type of communication is initiated by the Internet users themselves, for example, in social networks. This many-to-many communication reveals a significantly larger spectrum than the aforementioned types with regard to addressing and performance. On the one hand, individualized one-to-one messages and offers can be generated and sent. This results in the development of the new market segment **customer-to-customer** (C2C). At the same time, like and dislikes and/or offers are sent to a large circle of people known more or less well (e.g., via *Facebook*, *Google+*, or *Twitter*). This **hybrid use of social media** illustrates why above we spoke of a value adding as well as a value destroying impact of social media. This type of communication, which is largely independent of companies, represents a great challenge for them, because it involves a **reduction in or loss of the companies' information sovereignty**. Some companies are afraid of this loss of information and thus power. Yet in most cases it has long happened!

> ▶ **Food for Thought** During a presentation at the head office of a large **French retail company**, it arose that the managers there had ignored social media to a large extent until now. They were quite simply not taken seriously and therefore not monitored. For this reason, the top management did not realize that there were already 17 different fan pages about this company on *Facebook*—yet all of them with the tenor "I hate XY." And these pages had up to 60,000 fans. Therefore it is necessary that **Even if a company has not recognized the relevance of social media, maybe your own customers already have!**

In the course of Web 2.0, this is how **alliances** (by the likes of blogs, forums, and communities), who discuss with each other, work out solutions for each other, analyze them, and, where applicable, market them without a company being able to have direct influence. The vast number of private blogs, the increasing significance of social networks and media sharing platforms, as well as the time spent there make their increasing relevance clear. The original mass medium of the Internet has turned into a medium used by the masses even for one-to-one and group communication. Figure 1.13 shows how this has changed the way information is exchanged. The **classic linear communication** used to dominate. In doing so, companies were active as senders who sent out specific messages to predefined target groups via selected channels. This type is being increasingly supplemented or partially replaced by circular and polychrone (i.e., on various time levels) **communication**. Companies can still act as senders and send certain messages via selected channels to the target group. However, people—as part of the target group or irrespective of it—can then become active and send other contents via other channels to additional target groups. This **second focal step of communication** is conducted detached from, uncontrolled by and not controllable by the initially communicating company. With this in mind, it becomes more and more important for companies to achieve the "right" initial spark in communication, in order to make use of the viral processes partially setting in on behalf of the company's communication goals

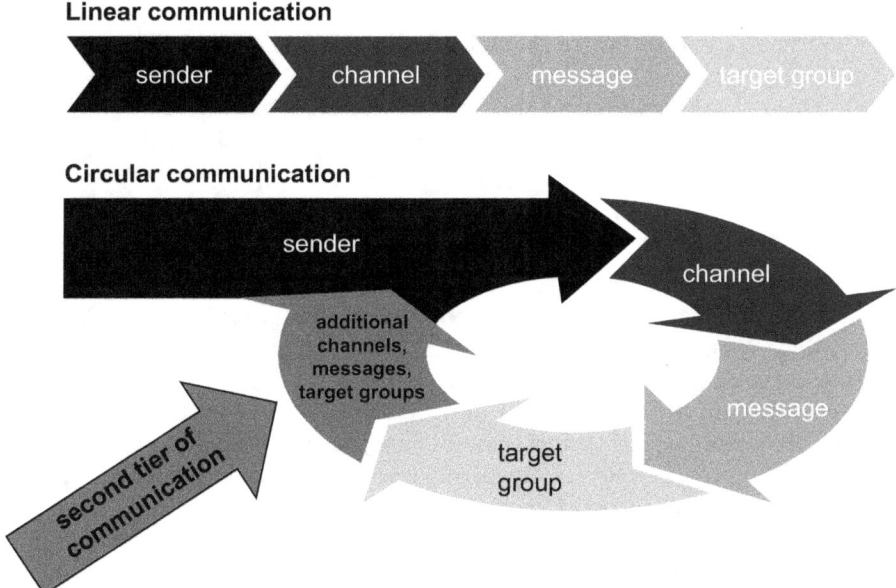

Fig. 1.13 Change of the communication process—from *linear* to *circular* (*Source* Author illustration)

(cf. article Ziems). In any event, it is imperative as little munition as possible is supplied for a so-called shitstorm.

Social media promote the **circular communication** displayed in Fig. 1.13, which can be characterized with the acronym **CIIS**:

- Collaborative (in the sense of the cooperation of the users in favor of or even not in favor of a company, a brand, or an offer)
- Interactive (in the sense of the users' interchange with each other and/or with the company)
- Iterative (in the meaning of repetitive, as, e.g., criticism, complaints, proposals, etc., are presented online until there is, in the user's opinion, an appropriate reaction)
- Simultaneous (in the sense of a simultaneousness of various strings and contents of communication)

Think Box
- Which forms of communication dominate in my company—one-to-mass, one-to-many, or one-to-one?
- Have we already found out whether we have arrived at the "ideal" point of individualization with regard to our earnings?

(continued)

> - What possibilities of the individualization of performance do we offer nowadays?
> - What developments among our competitors are foreseeable?
> - Who in my company could handle the core issues "individualization of communication" and "individualization of performance" more extensively and elaborate appropriate concepts?

If your company engages in the possibility of becoming more intensively involved in social media, then you should follow the **parable of social media marketing**. What is that about?

How do you behave when you arrive at a cocktail party? Do you shout out that you have also arrived when you get to the entrance? Do you offer your services (whether bicycles, bananas, skin creams, or financial investments) aggressively and for everybody to hear? After all, the party guests can't just run away, they have to listen to you. Yet probably not! So why do companies frequently act in this way in social media—and end up being punished by the target people with ignorance or "nasty comments"? How should we then go about it—in line with the parable of the cocktail party? You join in without attracting attention; go up to one of the tables and first listen to what the other guests are talking about. Then you join in by maybe asking a question or by making a contribution and slowly and appreciatively try to get the attention of the others. If, during the conversation, solutions are being looked for, for which you are predestined as a person or company, then information about your own offer is appropriate—not beforehand.

> ▶ **Food for Thought** Advertising messages in social media represent a disturbance!

When **integrating a company in social media**, attention must be paid to the companies and that their promotional messages are **not** initially **given a dominant role**. After all they are called "**social media**" and not "**commercial media.**" How company, marketing, and in particular sales goals can be achieved will be shown in the following chapters.

> ▶ **Remember Box** Social media must not be misinterpreted as another mere sales, advertising, or PR channel.

How can the challenges described be merged appropriately? By means of the **DiSoLoMo** trend that covers the dimensions **digital, social, local,** and **mobile** (cf. Fig. 1.14). In the first instance, the challenges for companies lie in the digitalization of products, services, and complete business concepts—as described before. An additional challenge is connected with **social media** because here a previously unknown power countervailing the established providers is emerging. The word "social" confronts us in increasingly more applications, from "social TV" to "social

Digital   Social   Local   Mobile

Fig. 1.14 The DiSoLoMo trend (*Source* Author illustration)

commerce" and "social plug-ins" down to "social CRM" (cf. Chaps. 2 and 7 for more details). In addition, the **localization of the users** and thus the **regionalization of demand and supply** will become more important. At the same time, an increasing tendency can be ascertained that location-based communities are also settling down in addition to location-based communication, which converge, for example, through the (spontaneous) check-in services described. Apart from that, the **mobile access** to Internet services by means of using smart phones and tablet PCs is increasing dramatically.

How can companies act in order to accommodate this trend during the **development of their own business model**? The **strategic gameboard** can provide an important overview of this (cf. Fig. 1.15). First of all, it poses the question of whether the company is operating on a market with **new or known rules**. In addition, the question is posed whether the **total market** or only a **niche** is to be served. Yet before we start off with innovations in the niche or even in the total market, we must know more precisely which new rules already apply in the market. We will therefore occupy ourselves with the impacts of the social revolution in Chap. 2 and demonstrate which basic needs fuel people in social media.

In doing so, it must be taken into consideration that the entire **corporate field** is changing dramatically at present:

- The **field is becoming larger**, because physical restrictions on goods and services, communication, and the demands of goods and services are becoming less important (e.g., due to zero gravity thinking and dematerialization).
- At the same time, **new rules of the game** are taking effect because, for example, more efficient performance-oriented invoicing systems are coming into operation.
- In addition, **new playground equipment** is being continuously introduced, as presented by the social networks (e.g., *Facebook, Twitter, Google+,* and *Pinterest*).
- Moreover, **millions of additional players** are rushing onto the field, because it is more or less possible for every Internet user to participate in any kind of communication with questions and own contents.
- At the same time, the **field is being extended to the 3rd dimension**, because the nature of the provision of information—based on **big data**—facilitates a three-dimensional CRM (cf. Fig. 1.6).
- If "**trust**" is introduced as a component relevant to action, the result is a **field extension into the 4th dimension** (cf. Fig. 6.4).
- In addition, the **speed of the game** is increased dramatically, because information is not only available in a density previously unknown, but the changes to it

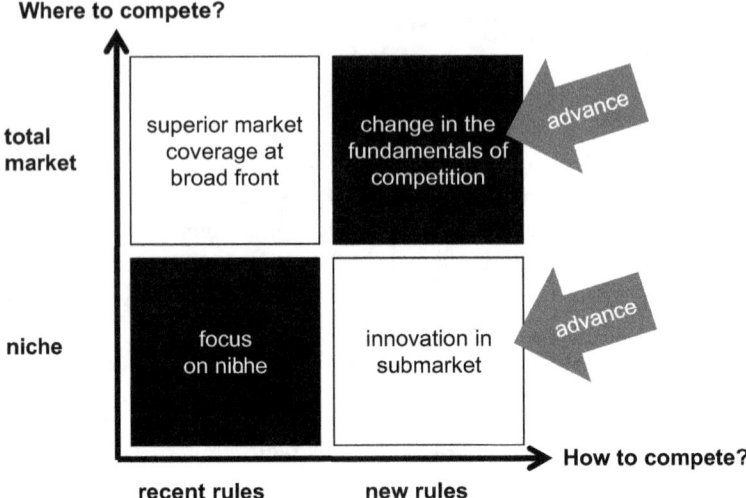

Fig. 1.15 Strategic gameboard—could we change the rules in the market? (*Source* Author illustration)

are often available in real time. This is where the necessity of real time marketing becomes apparent.

This entirety of change leads in some companies to a downright **state of shock**—not really a success strategy for mastering digital Darwinism! In the past, the credo "whoever moves, loses!" applied. Today we say: "Whoever doesn't move today has already lost tomorrow!" Yet when do we want to move as a company? Do we see ourselves as **first movers** or **fast movers** by actively seizing trends at an early stage? Or does our company rather fall into the group of **late movers**, who are happy to let others go first? In the light of the speed of change, the risk that the late moves shall become **first losers** is becoming greater! The **adaptability of our business models** is advancing to be a strategic competitive advantage!

> **Think Box**
> - Does my company use the new technologies as well as the gigantic flow of information to achieve competitive advantages?
> - Which competitors are one step ahead of us?
> - Are we in a position to change and, in particular, do we want to change the rules of the game on the market? Do we have the courage to lose sight of the familiar coast and head off to new destinations, trusting in sound navigation?
> - Do we dare to set out in only one market niche or are we ready to set new rules on the strategic gameboard on the total market, as well?

What do we need to do this? Courage and strength? And instead of a Chief Executive Officer, maybe even a **Chief Destruction Officer** instead, who is ready to break out of worn-down channels and to initiate an extensive restructuring of his own organization (cf. Fig. 1.16). This is why the request for a dance is "It's easier to kill an organization than to change it" (Peters, 1997, p. 71). You can read about important impulses on this concerning the digital transformation in Chap. 9.

One important **task of a Chief Destruction Officer** comprises shouting out to those in positions of responsibility in the company: Let us avoid the **marketing myopia**—i.e., a marketing shortsightedness that has already ruined many companies. There is no lack of statements saying that this marketing myopia is running rampant again nowadays. A recent example for that phenomena is the voting campaigns of the US presidential candidates Obama and Romney in 2012. Both nearly ignored the ethnic group of Hispanics, which represented one tenth of the US eligible voters and is expected to further growth. In average they spend only around 4 % of cumulative total investment on these voters. That behavior had and has an impact on the economy, because US companies ignore this ethnic group as well—a community expected to grow to 30 % by the year 2050 (cf. Llopis, 2012).

Another example also gives a good look on this phenomenon: the German public broadcaster *ZDF* underestimates the power of the **original channels of** *YouTube*. These are special interest channels financed by advertising and thus free of charge for the users, and which were introduced in 2012. The managers from *ZDF* stated: "These web-based platforms offers will have no impact on the TV market. The use of the Internet on TV appliances is too limited." The corresponding statement by the German public broadcaster *ARD* goes in the same direction: "New special interest channels are no competition for us. We will not be changing our focuses due to the start of the YouTube program" (Pohlmann, 2012). Even if the forecast of *Robert Kyncl, Google's* Vice President and Global Head of Business at *YouTube*, does not entirely hold true that in 2020, 75 % of all video contents will be transmitted via the Internet, the impacts on classic TV providers will probably be dramatic (cf. n. a., 9.10.2012, p. 6)—even if nobody really wants to see that now! But being in positions of responsibility, we have to! In the long run, a development is even foreseeable whereby *Facebook* will become a **personalized TV channel**—hopefully with no zapping and no spam!

It is better to accept such a **challenge**, which in the beginning could only endanger parts of the own **business model,** at an early stage and to respond to it. For especially advertising customers could appreciate the target group-specific

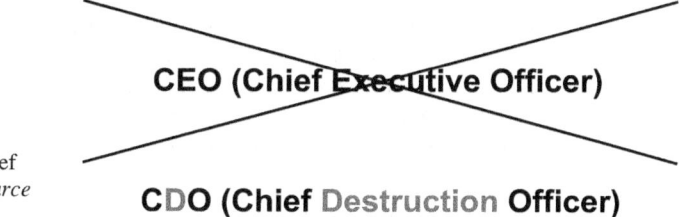

Fig. 1.16 From chief executive officer to chief destruction officer (*Source* Author illustration)

offers of the *YouTube* channel. And younger target groups, which already use *YouTube* regularly nowadays, will gladly make use of the additional offer, which they can access depending on their preferences regarding time, place, and contents. Due to the individual access to different clips, not only an **individual special interest channel** can be compiled. It can even be easily accessed on mobile devices such as smart phones and tablet PCs—not only on Internet compatible smart TVs. Thus there is a high probability that for more and more users, the classic **lean-back TV** will turn into **lean-forward TV**.

At this point it might be worth mentioning that the startup of one of today's well-known online retailers originally "only" involved one product—in this case, books. Nowadays, however, all universal and specialized mail-order companies equally feel challenged by this company. The provider is called *amazon* and was started up by *Jeff Bezos* in 1994/1995 "merely" as a special interest provider. And today? Companies such as *Walmart* and *Kmart* and others have continuously lost sales in the past 5 years. At the same time, *amazon* has achieved double-figured growth rates. *amazon* has "hijacked" sales as it were by transferring turnover from the physical world bound to real stores into cyberspace. The consequence? *amazon.com* has become the world's largest online retailer. The business models and with them the trading landscape change dramatically, thanks to new providers. And business models used to success for centuries bid farewell to the market, and new top dogs distinguish themselves.

How did that happen? Nowadays, the customer goes shopping, sees a new TV, scans the price tag with his smart phone, and does some research. In this way, the customer sees, for example, that *amazon* offers the television at a much lower price. Or there is a better recommendation for a similar device—keyword "social recommendation." If the customer trusts this (anonymous) recommendation more than the salesperson, he will walk back out of the store—here a significant **trend away from the physical world** and from the real shopping center **to the online store**. Stationary retailers have to face the challenge of **showrooming**: customers downgrade brick-and-mortar shops to mere showrooms—without buying there. That is a power that the customer is exploiting. He has extensive transparency over prices, delivery times, quality, and recommendations. And, in addition, his social network assists him in finding reasonable offers and suitable price-performance ratios. Many retailers still refuse to believe these developments. In doing so, there is an additional risk: retailers that cannot be found with their products during the online search in cyberspace do not exist for these customers in real life either!

The rudiments of a marketing myopia can also be found on the German **book market**. Although the stationary book retail trade again revealed a drop in sales of almost 5 % in 2012—in comparison to a further increase in online sales—the director of the *German Publishers and Booksellers Association* postulated at the opening of the *Frankfurt Book Fair 2012*: "An educated nation needs the stationary bookselling trade!" (Maier, 2012, p. 9). Even if one tends to agree with this statement from the bottom of one's heart, the majority of customers—as measured by their "lived" preferences for buying books—appears to have a different view. Not without reason are more and more well-known book stores that were successful for centuries being closed in central city center locations. The fact that in the USA 20 % of the book market is allotted to electronic books—compared with around 4 % in Germany—makes the challenge obvious. Even here, *amazon* sells in the meantime more digital books than hardcover editions.

Likewise on the **printed magazine and newspaper market,** dramatic changes are arising. The newspaper *Financial Times Germany*, for example, shut down its operations in 2012. The well-known US magazine *Newsweek* also made a grave decision due to a decline in circulation. As of 2013, there will be no printed edition. In the future—after 80 years of printing—there will only be an online issue bearing the title *Newsweek Global* (cf. n. a., 19.10.2012, p. 8). You can avoid a marketing myopia if you define the challenge like *Mathias Döpfner*, CEO *Axel Springer AG*, Hamburg: "Whether you think of a printed newspaper or a digital reader: In the future, the medium will become less and less important, yet the content more and more important" (cf. Baumann, 2012, p. 5).

Even on the **telecommunication market** there are massive upheavals. Increasingly more people—in particular the young ones—use the functionalities of *WhatsApp*. This is a cross-platform mobile message app enabling the exchange of messages without having to pay for text messages. The use of it is increasing rapidly and cannibalizes the turnover of classic telecommunication companies.

These are all **strong signals** for the changes becoming apparent for books, on the newspaper and magazine markets, and in the telecommunication industry. One is tempted to call out: listen to the signals! Or, when speaking to these companies: can't you hear the signals?

> ▶ **Food for Thought** The saying of the *Dakota Indians* still applies: **"If you discover you are riding a dead horse, get off."**

Yet how do many companies react in such a situation—and us maybe too?

- We substitute the rider.
- We call a meeting to analyze the dead horse.
- We reclassify the horse from "dead" to "not suitable for life."
- We harness several dead horses together so that they become faster.
- We squeeze in a training unit to learn how to ride dead horses better.
- We rewrite the requirements of all horses.
- We buy a stronger whip.
- We declare that, when dead, our horse is "better, faster, and cheaper" than that of our competitors.
- We reduce the standards so that we can continue to employ the dead horse.
- We visit other places to see how others ride dead horses.
- We buy people from elsewhere to ride the dead horse.
- We conduct a productivity study to determine whether more light-weighted riders can ride dead horses better.
- We appoint a committee to reanimate the dead horse.

- We declare that the horse does not need to be fed, causes fewer overheads, and thus makes a greater contribution to achieving the company's goals than other horses do.
- We say: "That's the way we always rode the horse."

Does the following guiding principle maybe apply here?

▶ Life is too short to leave it up to reality!

And yet, nowadays, it is very simple to identify **important signals**. An important source for this is *Google Trends*. It can be used to compare several websites with each other. It is easy to recognize that the **pattern of the universal mail-order company** *otto.de* dramatically resembles the supplier *neckermann.de* which has dropped out of the market in the meantime. The differences to the current winner *amazon.de* are striking.

Another task a Chief Destruction Officer has is to avoid a **regional marketing myopia**. Anyone who thinks that the USA is the **pioneer when it comes to social media** is terribly mistaken. It is **China** in particular, where the social media revolution is becoming notable—driven by the possibility of developing an online communication intensity that was previously not possible. It is thus essential to turn our attention to China in particular with its approx. 60 million app users of Baidu Tieba and around 36 million mobile users of *Renren*—in December 2013 alone (cf. China Internet Marketing, 2014). Apart from that, there are already a billion users of mobile phones about there: the smart phone becomes the "first screen" and ousts the TV to the ranking as "second screen." But even **South Korea** and **Japan** should not be ignored if we want to track down exciting trends and new applications.

The **global diffusion process of innovations on the Internet** is headed by countries like China and South Korea as **innovators** (cf. Fig. 1.17). During the further diffusion, the USA frequently belongs to the group of **early adopters**, who seize such developments relatively fast. Europe and the companies located here are frequently found in the **early or late majority** or even in the group of **laggards**. The spearhead of innovation can definitely be found in Asia!

Let us get back to social media marketing. Is there also a marketing myopia when dealing with social media? How does the **top management** react towards the **challenges of social media**? *Michael Buck* has made up a list of the top ten excuses for this which, together, form the **Management Excuses Hall of Fame (Shame)** (cf. Fig. 1.18; Buck, 2013).

Some of the reasons stated here are even reflected in the BITKOM study on the use of social media in companies. It is fascinating that in the **reasons against social media activities** mentioned in Fig. 1.19, besides the "**we don't reach our target group**" and the "**legal uncertainties**," the reason "**social media does not fit with our corporate culture**" was stated by an amazing 45 % of the companies. These companies respond to the relevance of social media from a sender perspective—and

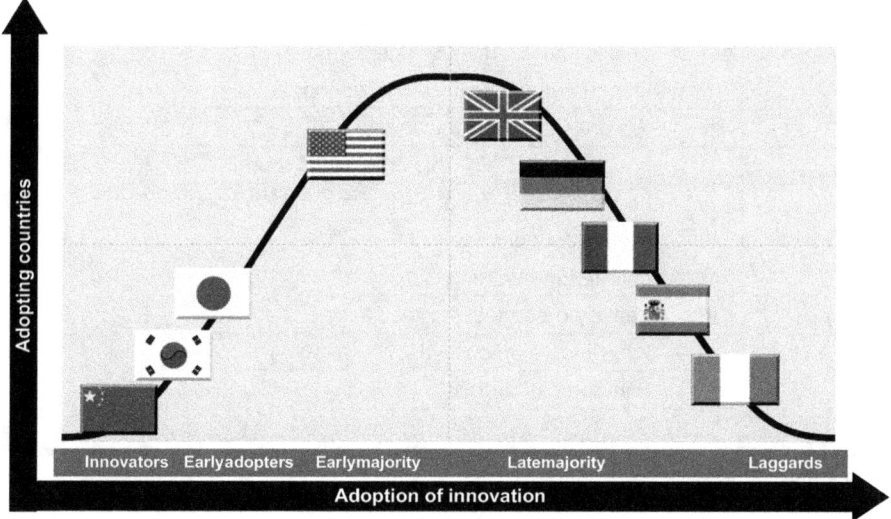

Fig. 1.17 Global diffusion process of innovation on the Internet (*Source* Author illustration)

not from a recipient perspective. This makes it clear that the anchorage of social media in companies must be accompanied by **change management** (for more details, cf. Chap. 9).

> **Think Box**
> - How strong is the presence of marketing myopia in my company?
> - In which areas and on what management levels is it particularly distinctive?
> - How can we reduce this dangerous shortsightedness? What could help us to do this?
> - Which countries do we orient ourselves towards when dealing with digital innovations?
> - Do we have an eye on the real innovators or are we settling with the early adopters of the "early majority"?
> - Do we also have a "top ten of excuses" regarding social media?
> - How can we take off the blinkers in this respect?
> - Who could become the "driver" of this?

Let us all take hold of a magnifying glass to counteract the revealed **marketing myopia**. In doing so, we will not like everything we will see. However, we can only decide how to deal with challenges if we recognize them at an early stage and brave them. Otherwise, the challenges will rather deal with us—and we will have no more alternatives!

Rank	Statement
1.	This is just a fad – it will go away!
2.	Our customers are not on *Facebook*!
3.	We only have B2B business!
4.	You work for *Dell* – this only works there!
5.	I cannot control what people are talking about!
6.	What about privacy?
7.	I cannot measure it!
8.	I have no budget!
9.	I don´t have time for it!
10.	I am too old for this!

Fig. 1.18 Management excuses hall of fame (Shame) regarding the use of social media (*Source* Author illustration, data source: Buck, 2013)

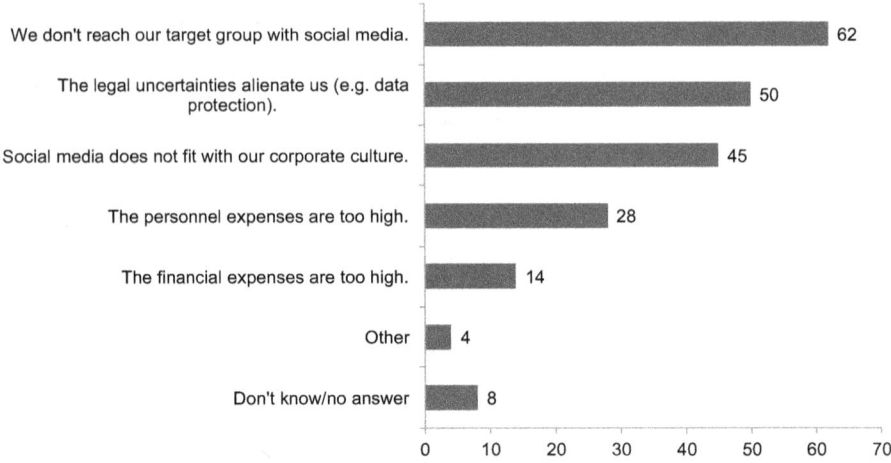

Fig. 1.19 Reasons against social media—in % (Question: "What are the reasons why you decide against the usage of social media for your business?"; multiple replies possible; $n = 391$) (*Source* Author illustration, data source: BITKOM, 2012, p. 22)

And no, the key powers behind **digital Darwinism** and also **social media** will not "pass by" like a wave of influenza. They will stay! And here, too, what *Aldous Huxley* so aptly said holds true: "You cannot dispose of facts once and for all by ignoring them." The **interchange via social media** after the telephone and email/letter will become accepted as the **third column of communication instruments**—in the course of the years even across all target groups.

For this reason, the following holds true:

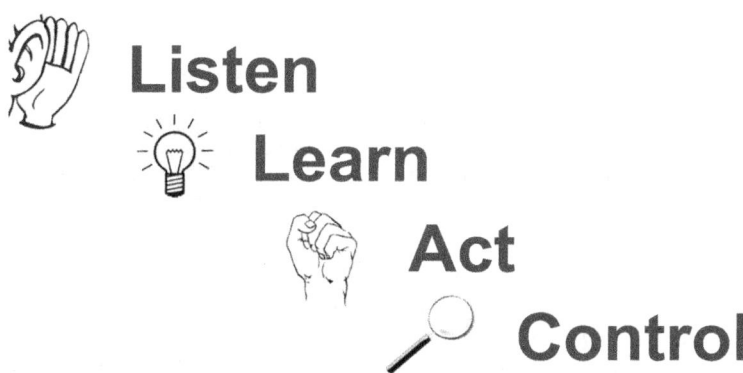

Fig. 1.20 Guiding principle for one's own actions—not just for social media (*Source* Author illustration)

▶ If you can't beat it, join it!

This book is meant to convey significant impulses to this end. The first important step is to learn to **listen** again. Companies were caught up in **send mode** for too long—and stuck to it! This was and also frequently still is the case when response quotas to emails, e-newsletter, mailings, classical response ads, and also to informational material in social media decrease dramatically. The reasons behind the rejecting attitude of our target persons are still not being sufficiently analyzed. For this reason, a concept of procedure is recommended at this point in time, which has proven successful as a **general guiding principle within the company**—but also on a private basis: it covers four steps **Listen, Learn, Act, Control** (cf. Fig. 1.20).

A good conversation always begins with appreciative listening in order to gather needs, interests, and moods. With **Listen,** the first step is to also approach social media (cf. the parable of social media marketing already mentioned). With that, particular empathy is to be displayed in order to get a deep understanding of the customers. The second step **Learn** is about recognizing the relevant correlations, interpreting behavior patterns, and developing methods of resolution. After **Act,** the implementation—and this step is frequently not implemented consequently enough—follows imperatively the **Control** phase. This is the only way the setup of a learning organization can be successful. And we will be learning a lot faster in the months and years to come, faster than was the case in the past centuries.

> ▶ **Remember Box Every good conversation begins with "Listen"!** And the basic principle of dialog marketing of test, test, test, which has been valid for centuries, also applies to online marketing in general and to social media marketing in particular. Nothing is as unstable as user preferences. That is why it is always good to listen, to continuously try out new approaches to win over the users' interest and keep it!

> **Think Box**
> - What does my company understand by dialog? Sending?
> - Are the systems of my company oriented towards dialog at all?
> - Are our employees paid for listening or for sending?
> - How do we reward employees that transport valuable contents "from the outside to the inside"?
> - Do such employees or the information they have gathered find an attentive ear on the "upmost management level"?
> - Or do effective filters screen off the "monsters of life" from the upmost management level?
> - Who would be able to break through such processes of blocking information as well as contaminating information?

When mastering the tasks lying ahead of us, let us follow the phrasing from *Goethe's Faust*:

> ▶ A mind once formed finds naught made right thereafter. A growing mind will thank you evermore.

Let us simply be a growing mind, which faces the tasks lying ahead of us with an open heart and an alert mind. Then the emerging changes will become challenges from which we and our company can equally learn and grow.

One thing is certain:

> ▶ The digital rat race has already started!

> ▶ **Food for Thought** "The process is evolutionary, you hardly notice it. But the impacts are revolutionary."

Wolf Bauer, Chairman of the Board of *Ufa*

Quick Wins

-
-
-
-
-
-
-
-
-
-
-
-
-
-
-
-
-
-
-
-
-
-
-
-
-
-
-
-
-
-
-
-

Digital Darwinism and the Social Revolution: What Basic Needs of Man Represent the Fuel of the Revolution on the Part of the Customer?

Today, we can justifiably speak of a **social revolution**, because the possibilities of the Internet help provide millions of users a previously unimaginable attentive ear and thus a **plentitude of power**. The central lever of the new power the customers have is their **worldwide network**. However, this is where we have to be careful not to let the **power users**, who dominate social media with their actions and opinions, dominate us in our actions and reactions. No matter what, it is, however, the companies' responsibility to establish an effective online **counterprevailing power**, in order to not end up being the chased in social media.

To understand the customer's side of the social revolution, we first need to understand the basic need structure of human beings, in order to be able to attend on them (cf. Fig. 2.1). Taking the own "ego" as the starting point in this **basic need structure**, the needs for solidarity on the one hand and freedom/autonomy on the other—facing each other in the area of conflict—are aspired. **Solidarity** covers the need for safety as well as the sense of belonging to a partner, family, group, team, and company. An attempt is made to be part of "something larger." The negative characteristics of this are dependency on third parties and self-abandonment. In contrast to that, there is the pursuit of **freedom/autonomy**. In combination with that, power and control are pursued. The uncontrolled satisfying of this need can lead to loneliness.

A further basic need of human beings is the quest for **creativity/self-fulfillment**. This is about creating something and to effect performance. An excessive expression of this need can lead to an inability to cope. Somewhat isolated from these three basic needs, we have the quest for **being**. This is about being accepted because you "exist"—not because you have done something. In an ideal situation of being, flow sets in. Then you feel in accord with what you are currently doing. The physical challenge experienced and your own capacity are in absolute balance. Only the moment counts—and time flies by unnoticed. A dominating factor when expressing this need for "being" may, in contrast, be the feeling of boredom and uselessness. The entirety of these four needs, aspired by every human being in varying intensities, represents the important drivers of human behavior.

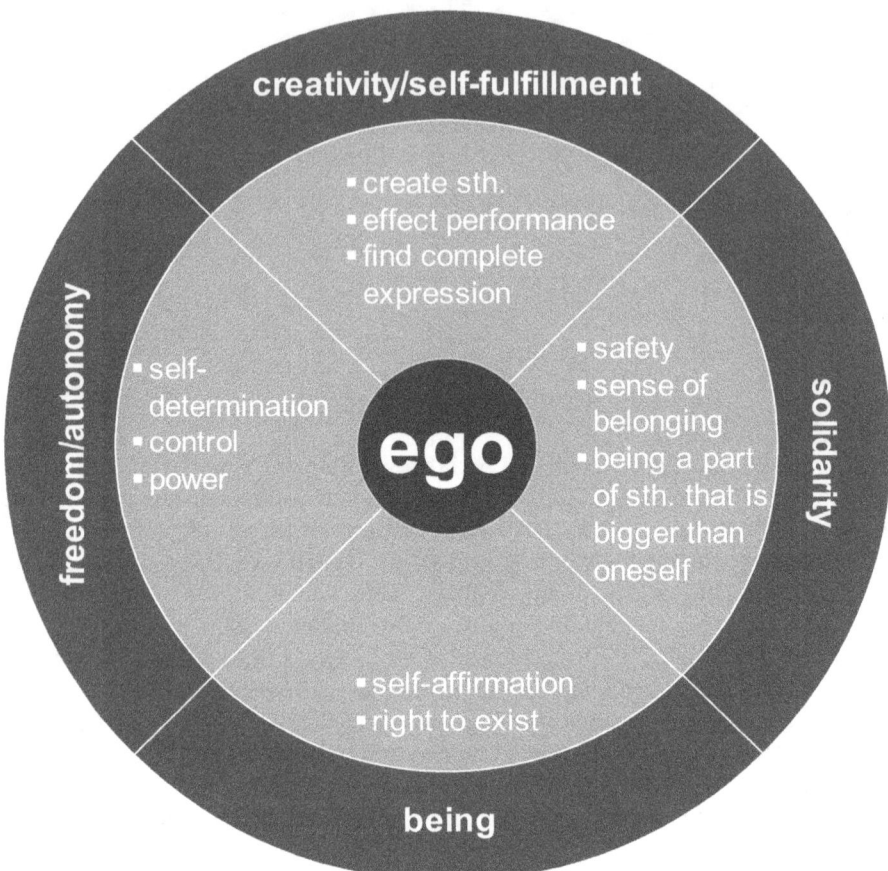

Fig. 2.1 Basic need structure of human beings (*Source* Author illustration)

Derived from these fundamental needs of human beings, we can now proceed in search of the **motives for specific behavioral patterns on the Internet** (cf. Fig. 2.2). The **motive** represents the reason for why we human beings do something. In the case of the target group that is of particular importance to us, that of potential customers and customers as online users, we must differentiate between commercial and noncommercial motives for what they do. **Commercial motives** include, for example, the attempt to purchase products or services for as low a price as possible. This motive leads people visiting price comparison sites (such as *bizrate.com, nextag.com, shoplocal.com, slickdeals.net;* cf. Ebizma, 2013). Commercial motives also lead to people selling their own services or other products (e.g., photos, videos, music recordings, texts, and hand-made crafts), without actually being a professional seller in e-commerce. The previously mentioned platform *Etsy* but also more diversified providers like *eBay* or *amazon* offer these semiprofessional sellers interesting means of access to the market. These help to

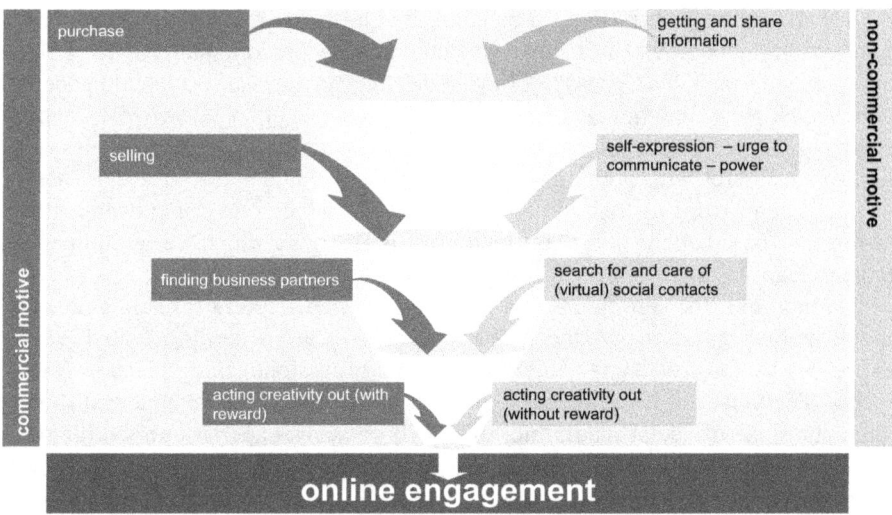

Fig. 2.2 Motive structures of online users (*Source* Author illustration)

make contributions towards satisfying the needs for autonomy and in particular for creativity. Apart from that, the search for business partners (also as a potential employer) can be promoted by networks such *LinkedIn, talkbiznow.com,* or *meettheboss.tv*. These activities based on the aspiration for solidarity.

Besides that, there is a large number of **noncommercial motives**, which result in increasingly more time being spent online. The dominant driver behind that is the quest for affinity. The basics for this are **getting and sharing information**, which is supported by online research using search engines and can make a contribution to commercial goals. The evaluation of or the participation in evaluation platforms, blogs, forums, and communities satisfies the need for affinity. In addition, the latter concepts offer the possibility of making one's own contributions, whereby the motives of **self-expression** and the **urge to communicate** as specific characteristics of creativity are accommodated. The entirety of these motives leads to millions of *Facebook* members elaborately creating their profiles and pinboards, updating them on a daily basis, and filling them with videos, photos, and/or recordings. The spectrum for living out one's own urge to communicate ranges from the answer to the simple question "What am I doing," which is answered via *Twitter* and *Facebook,* to the editing of complex topics on *Wikipedia*. In the case of the latter involvement, solidarity with the knowledge community is established. At the same time it can be ascertained that "Private television gave everybody the chance of 5 min of fame. In the mass culture of the Internet, the fear of losing significance in the ocean of opinions opens the private locks" (Thiel, 2012, p. 27). Is a lot of what is happening here a cry for attention and a scream for affinity in a world where more and more securities are disappearing?

A **striving for power** can also be connected to the urge to communicate based on freedom/autonomy. Until now, consumers were predominantly a part of an unorganized, invisible "mass." They were hardly able to exert influence on companies. Now, individuals can vote via the Internet in social networks and by means of blogs, forums, and communities, put the fear of God into companies. This is also successful with politicians when diligent researchers provide evidence of neglect when writing up their dissertation. All in all, a previously unknown **position of power**—based on the **power of the mass**—is emerging which we, as companies, have to learn to deal with.

Through this, the **significance of previous opinion makers** is being embedded in the public. Until today, the (public) opinion was primarily affected by the exposures by (professional) opinion makers in discussions taking place in (mass) media. The voices of the (alleged) experts dominated. Due to the greater diffusion and active use of social media, not only will previous experts find it difficult to convey their monopolies of opinions, but a much greater diversity of opinions will emerge. And the job of our companies will be to catch them in the organization (cf. Chap. 9).

Specific offers on the web enable users to live out the quest for affinity, creativity, and autonomy simultaneously—and at no risk. Offers such as *FarmVille* or *CityVille* make it possible to start up dream careers and to take over the power over one's own company—at least in the virtual world (cf. fliplife.rtl2.de). The objective of such offers is to accommodate the quest for social recognition by becoming, for example, a boss or by winning an international sport tournament—only in the virtual world, however!

The motive of **acting creativity out** can, on the one hand, refer to the further development of third-party performances. These include, for example, the involvement in the development of *Lego* bricks as well as the answering of complex research questions which, for example, are presented by companies such as *Novartis, DuPont,* and *Procter & Gamble* on *innocentive.com*. In some cases, in doing so, commercial motives may play a role if creative services are also rewarded by the companies. The creativity motive may, on the other hand, refer to the design of third-party performances, which the customer himself wishes to buy. This is the case with the concepts of *spreadshirt.com* or *cerealize.com*. Contributing one's own creativity can at the same time make a contribution to the motive of **self-expression** (the need to be) in the relevant reference groups, whereby the **search for and care of (virtual) contacts** can be supported as expressing the quest for affinity. For many users, this represents the main motive for their involvement in social media.

This **analysis of the needs structures** and the **motive structure of online users** provides an important background for actions for the organization of our marketing and communication measures. For, only the extensive consideration of the insights presented safeguards what is vital for an involvement of our target segments.

▶ Relevance!

All of what has been previously said can be summarized into the following **relevance triggers of an engagement in social media** from the users' point of view:

- Entertainment/fun
- Education/growth/enable people
- Save money
- Save time
- Solve something

It becomes obvious: we definitely have to offer the users **added value for the engagement in social media** to motive their involvement. **Relevance** will only be created if at least some of the added value dimensions stated here are present. Concerning our own dedication to social media, we can take the *Facebook* mission as a reference:

Give people the power to share and make the world more open and connected.

Think Box
- To what extent is the need structure of the target persons taken into consideration in general when organizing our marketing concept?
- Which needs do we address in particular with our social media involvement?
- Which needs should and can we address?
- Which triggers of an involvement can we apply in a particularly credible way: entertainment/fun, education/growth/enable people, save money, save time, or solve something?
- Is it being monitored whether we are applying or can apply such triggers in a really targeted manner?

At the same time, we must recognize what **expectations potential customers and customers** have of us as a company—which we should take into consideration when organizing our products and services. To put it simple, these can be characterized by the buzzwords "**me, all, immediately, and everywhere.**" How these expectations are defined in the individual cases is shown in Fig. 2.3.

Due to the great intensity of competition in almost all sectors, under "**me**," the customer can take the liberty of expecting or even demanding great **appreciation in the interaction** among other things. If he is denied this, there are usually many competitors waiting to win over the customer. "Lived appreciation" includes a **correct personalization**, for example, the correct addressing of the target person by name. At the same time, the expectation increases that, as an individual, he is

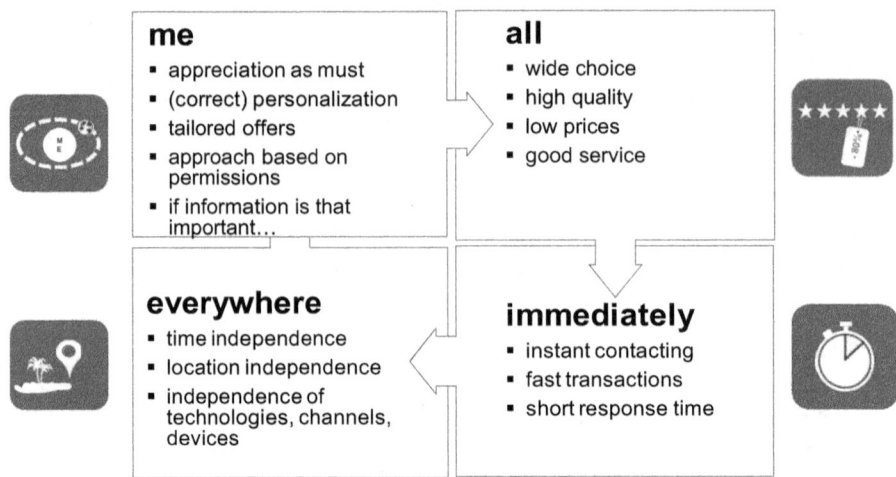

Fig. 2.3 Ascertainment of the new expectations on the customer side (*Source* Author illustration)

taken seriously with specific and, under certain circumstances, with customized wishes and thus receives personalized communication and offers. The **demands from the potential customers' and customers' perspective**—which we see ourselves increasingly confronted with—are summarized, for example, in the following questions:

- Are the offers customized to my needs?
- Does the sender speak my language and does he put that across?
- Do I receive precisely that information that I wish by email, e-newsletter, mailing, posts, status updates, etc.?
- Are my instructions for addressing me by mailing, posts, telephone, email, and fax respected by the company advertising?
- Can I find the information I want fast both online and offline?
- Can I simply order and pay?
- Do I find the help I need?
- Can I effect the desired transactions when I choose to and from where I choose to (research work, reading newspapers/magazines, posting inquiries, and placing orders)?

The question of appreciation is anchored in another point—**the way the terms** (General Terms of Business) or other agreements **are dealt with**, above all data policies presented to the potential customers and customers. The terms of use on *Facebook* are laid down on three pages. A further 12 pages cover the "Statement of Rights and Responsibilities" of *Facebook* and a further 20 pages define the Data Use Policy (as of 2014). The most frequent lie on the Internet is surely: "I have read the Terms," as they are frequently very lengthy and presented in small writing. Despite or rather due to this behavior of our customers, we as companies are well advised not to plant any biased, uncommon, or surprising clauses and conditions on the users. Either they may be illegal and thus become ineffective or they are legally agreed upon and are interpreted to the customer's disadvantage. If the conflicts involved are dragged out in social media, long-term damage to one's image can be the consequence. Both impacts should be avoided.

The expectations regarding **individualization** materialize in the **granting of permissions** to contact. These permissions are specific authorizations a potential customer or customer gives a company regarding the "permitted" ways of contacting them (such as by letter, email, telephone, and/or fax). The permissions called **tokens** needed to access user data on *Facebook*, for example, are of particular importance (cf. Chaps. 6 and 7). These permissions can be withdrawn by the potential customer or customer at all times. Companies are obliged by law to strictly observe these permissions for contacting.

The self-centeredness also materializes in the **demands on entertainment electronics**. Fewer and fewer—in particular younger—people want to watch films and documentaries when they are actually being broadcast on TV. With that said, **on-demand and streaming services**, which besides linear TV are also obviating the need for CD and DVD players, are booming. The manufacturers of hardware will lose in favor of content sales platforms (such as *amazon*, *Spotify*), as long as they themselves do not become content marketers (such as *Apple* with *iTunes*).

An additional element of the self-centered expectancy materializes in the statement: **if the news is that important, it will find me** (cf. Mathew, 2008). This means that people believe that they no longer need to actively search for information and offers, because they are presented with it in particular via posts in social media—if relevant!

The customer becomes a **master of communication**. He decides when, who, about what, and via which channels one can communicate with him. To this end, the customer is granted this power due to his legally protected position and the support coming from social media.

The expectant attitude "**all**" shows the great expectations customers reveal in most sectors. Customers have learned that "everything is possible" frequently applies:

- Award-winning products of the *Consumer Reports* at *Walmart* are to be found in the range as best buy products.
- *H&M* offers clothing by the designers *Jimmy Choo, Roberto Cavalli,* and *Sonia Rykiel* at favorable conditions.
- *amazon* offers a wide and extensive range, combined with a highly personalized recommendation of other "matching" products—including delivery the following day.
- Increasingly more brands are enabling the individualization of the product—from *Ray Ban* spectacles over pullovers by *Laura Biagiotti* to the *Prada* bag. In the case of some prestige brands, anyone can now become a self-made designer.

- On the Internet, an almost inexhaustible range of information can be found—around the clock, frequently highly topical or as a news stream even in real time and mostly free of charge.

At the same time, an **"it won't be paid for"** or **"as long as it's free mentality"** is prevailing in many areas of the Internet. That makes it very difficult for companies to motivate users to overcome **paywalls** they have installed to earn money with their content on the Internet. The acceptance of paywalls by the customers still represents a great challenge for the majority of publishers. The user has learned that he doesn't have to pay anything for great tips—for example, where there are savings at the moment (such as at *tightwad.com*), which hotel is offering a particularly good price-performance ratio (such as at *trivago.com*), which airline offers the cheapest flight from New York to Delhi (at *farecompare.com*), and for all possible and impossible answers at the site from moms for moms at *soulmom.com*. And there is another offer of news—the *Huffington Post*. Why should he then pay for editorial news in the *New York Times*?

This is how many users consistently avoid overcoming the paywalls the publishers—called **content providers**—have set up or are currently doing so. Again and again, attempts have been and are being made to put a price tag on the contents also offered on the Internet with great effort. Whereas few people would come up with the idea of asking for a free copy of a printed newspaper at a kiosk, free access to the same material in an online format is quasi-presupposed, a fact that can lead to existential problems for many companies. It will thus be interesting to see how the **fee-charged apps** of selected publishers (at present, e.g., of the newspapers *New York Times* and *The Financial Times*) will be accepted by user groups in the long run.

For, we as users need to be clear about a great **risk of the "it mustn't cost anything mentality" on the Internet**: "Anybody who is not willing to spend good money on good journalism, puts himself at the mercy of commerce and the search engines, which are ready to pounce on our data. And when the last decent newspaper has disappeared, all that remains is gossip" (D'Inka, 2012, p. 1).

The expectant attitude of **"everywhere/immediately"** is also being fulfilled by many companies in the meantime. Mobile availability—not only via telephone but also as access to the Internet—is taken for granted nowadays in the developed industrial nations. The buzzword for this is **always on**—meaning "always being available"—irrespective of whether this takes place in a work or private environment and stationary or mobile. This also causes the thresholds between private and public or professional spheres to increasingly merge. For this reason, potential customers and customers frequently expect around the clock access to the customer service center: every day, 7 days a week, 365 days a year—without giving the financial implications for the company a thought.

The access to company offers is thus increasingly shifting from "classical business hours" at certain places to a **customer-driven interaction process flexible in time and location**. Thus, the potential customers and customers are able to both receive and send from anywhere and around the clock. This challenge makes great demands on companies. In this **instant society**, the motto of "any channel, any device, anywhere, anytime" holds true.

In addition, there is the expectant attitude "**immediately**," which causes **acceleration effects** in various areas. Generally speaking, potential customers and customers give companies increasingly less **time to respond**. If there is still no response to an email after 4 h, this is followed up many times. And why should a customer wait 2–4 weeks with a seller if an order placed with *amazon* is usually fulfilled within 24–48 h—or even the same day? Past experience with *amazon* is drawn on as a benchmark (i.e., as a reference value) for the rating of other companies—even beyond the borders of sectors. Whether this appears appropriate in a specific case from a seller's point of view is of little interest to the I-driven customers. **Channel hopping** helps the user to punish the slow companies by means of a mouse click on a competitor—possibly with the consequence that the slow company has lost the user forever.

This "immediately" leads to a further interesting development, which can be described by the term of a **culture of now**. It is not only increasingly ascertained from younger target groups that when watching TV (which in many cases takes place on a laptop, smart phone, or a tablet PC), they not only check their *Facebook* account or email inbox on a regular basis but also send text messages and compare notes about various topics via *Twitter* or *WhatsApp*. **Multitasking** causes employees and managers to continuously check their email inbox on their *Blackberry*, *iPhone* & Co.—even and especially during ongoing conferences, meetings, and lectures. Recent studies on multitasking make it clear, however, that this is not the nature of man and one therefore achieves significantly worse results when concentrating on several tasks at a time. In addition, it becomes clear that this phenomenon not only holds true for men, but that women are also not really able to multitask (cf. DGUV, 2010)! But what is the greatest **driver of the culture of now**, if it is not an improvement in service provision?

▶ The danger of missing out on something!

When it comes to **time as a critical success factor**, the process presented in Fig. 2.4 thus needs to be considered. When examining an offer, a customer's **motivation** first increases. At the highest point, a request for more information or an order is frequently made. After that, the level of motivation goes down again because other offers are fighting for attention and one's own doings slowly fall into oblivion. The faster the requested catalog, the first e-newsletter or sample copy of a newspaper or the delivery of *Zappos* arrives, for example, the greater the motivation to be found, which can have the effect of being promotional towards the image and ordering. One thing is certain: the later the request happens, the less remains of the initial motivation, because other companies have possibly—and faster—provided further information or submitted interesting offers. And with that, we are possibly not talking about weeks, but days and—in particular in e-commerce—hours!

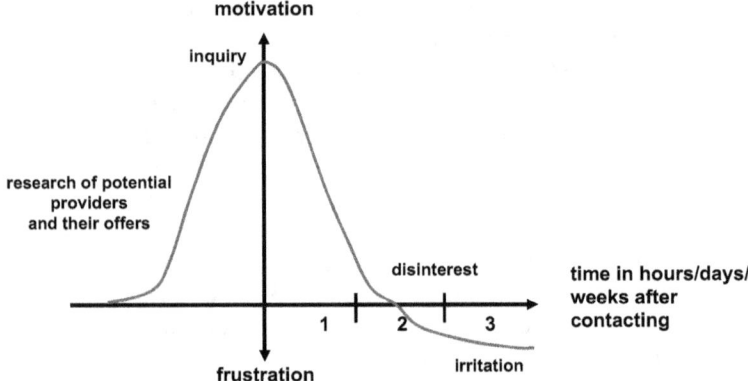

Fig. 2.4 Speed as critical success factor (*Source* Author illustration)

If, in contrast, the information or the offer requested or the goods ordered (online) arrive significantly later than expected, the delivery is possibly met with **disinterest**, because the potential customer no longer remembers his request or has decided in favor of a different seller in the meantime. However, **irritation** may also arise, because the information or products were not available at the expected point in time and one does not feel having been taken seriously enough as a person or company. As a result, information remains unread or goods are returned.

> ▶ **Food for Thought** "With customers today being increasingly connected, informed, and ultimately empowered, their expectations only escalate. In short, they are more discerning and demanding than ever before."
> *Brian Solis*, Principal of the *Altimeter Group*

Think Box
- How thoroughly have we already looked into the requirements "Me, all, everywhere, immediately"?
- What are the consequences of the expectations for our value chain?
- How well positioned are we compared to competitors?
- Which levers can be pulled promptly in order to meet the expectations, for example, of accelerated internal processes?
- Where is the corresponding responsibility to be placed in my company?

With that said, the question arises as to how the **customer journey**—meaning the "customer's journey to the company"—and **customer expectations** have changed it. This "journey" covers the various phases a customer goes through before deciding to purchase a product or service. At the same time, particular

```
┌─────────┐  ┌──────────────┐  ┌──────────────┐
│Stimulus │  │First Moment of│  │Second Moment of│
│         │  │Truth (POS)   │  │Truth (Experience)│
└─────────┘  └──────────────┘  └──────────────┘
```

Fig. 2.5 Classical sequence: stimulus–FMOT–SMOT (*Source* Author illustration)

Fig. 2.6 Classical AIDA formula (*Source* Author illustration)

significance is attached to the so-called customer touch points of the company or a brand, with which the customer comes into contact on this journey. Some facets of the classical purchasing process have shifted due to the entering of the online era. Until recently, only the first and second moments of truth were differentiated after the stimulus in the course of the decision-making process to purchase (cf. Fig. 2.5).

The **first moment of truth** (**FMOT**) describes the moment when a potential buyer can physically examine a product or service for the first time. This is where the expectations established by advertising, meet the "harsh reality" of the product or service. The **second moment of truth** (**SMOT**) covers the moment when the buyer actually makes use of the product or service. This, in turn, is where the expectations established upon first inspection are in contrast to the actual performances and experiences when using the product or the service. One speaks of the "moment of truth" because these two "moments" reveal whether the expectations formed in particular by advertising, the presentation of the offer, or possible advice at the POS is also actually fulfilled. It was possible to present this traditional customer journey in a concise and very simple way with the classical **AIDA formula** (cf. Fig. 2.6).

However, we should say farewell to this classical concept because, at present, a fundamental change is taking place in the decision-making and purchasing process of the customers. The first and second moments of truth are now coupled with the **zero moment of truth** (**ZMOT**) in the online era (cf. Fig. 2.7). This means in particular the online access to an almost unmanageable amount of third-party information preceding the other two "moments." Part of this so-called user-generated content is reports from other people providing information about their experiences prior to, during, and after the act of purchasing and using.

The information from blogs, communities, comments on *Facebook*, *Pinterest,* or via *Twitter* enable potential customers "**to help themselves to third-party experience**," which shapes the content of this ZMOT. With that, own possible experiences

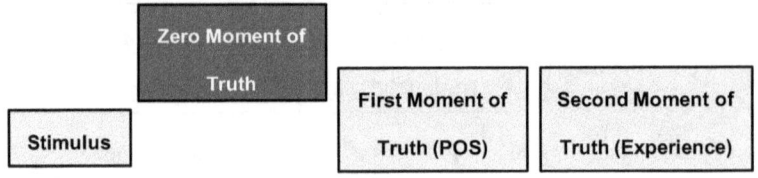

Fig. 2.7 Position and sources of the ZMOT (*Source* Author illustration)

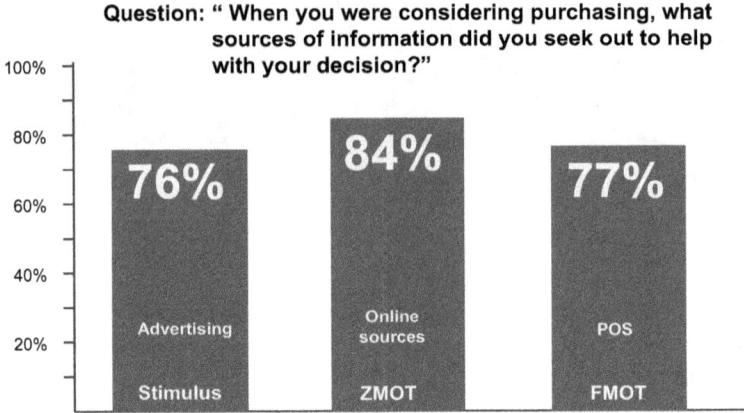

Fig. 2.8 Importance of different sources of information in the buying decision process—in % (US consumer, $n = 5.003$, 2011) (*Source* Author illustration, data source: Lecinski, 2011, p. 19)

are frequently "anticipated" by unknown third parties by accessing reports, photos, and videos. A large amount of information can thus give an impression about the pre-sales, sales, post-sales, and usage stages of other people even before the potential buyer has been able to get his own impression of the target object. The ZMOT is thus fed by the experiences others have had along their **customer relationship lifecycle** (for more details cf. Kreutzer, 2009, pp. 49–56; cf. Chap. 8).

When asking oneself how important the **significance of the ZMOT** is for companies, the following can be determined. According to a study of US consumers in 2011, the number of sources of information availed of increased from 2010 to 2011 from 5.3 to 10.4 (cf. Lecinski, 2011, p. 17). At the same time, Fig. 2.8 illustrates what significance is attached to the ZMOT nowadays during the buying decision process.

> **Remember Box** Classical Recommendation Marketing Is Experiencing an Unforeseen Renaissance!

And a further result deserves our attention: "It's well known that consumer research expensive products like electronics online, but coming out of the recession, consumers are more scrupulous about researching their everyday products

Fig. 2.9 Consumer trust in types of advertising and promotion—in % (multiple answers possible, around 60,000 US online adults) March 2013; % rating 4 or 5 on a scale of 1 (do not trust at all) to 5 (trust completely) (*Source* Author illustration, data source: Forrester, 2013)

such as diapers and detergent, too. More than a fifth of them also research food and beverages, nearly a third research pet products and 39 % research baby products, even though they ultimately tend to buy those products in stores, according to WSL Strategic Retail, a consulting firm" (Byron, 2011, S. 1). This makes it clear that a new target group is becoming more important:

▶ ROPOs: research online, purchase offline

Why are companies better off recognizing the **relevance of the ZMOT** and to react accordingly? The most powerful arguments for this are provided by Fig. 2.9. In response to the question about which **sources of information** the customers **trust most**, "**brand or product recommendations**" are—not surprisingly—in the first place with 70 %. It is interesting that "**professionally written online reviews**" and "**consumer-written online reviews**" with 55 and 46 % follow in second and third place (cf. Forrester, 2013). It must be pointed out, however, that according to a recent study, about every third hotel rating on the Internet is a fake. Yet it was determined that merely one single rating can increase the number of bookings by 70 % a year (cf. Oberhuber, 2012, p. 45).

It becomes obvious that consumers trust statements made by "professional authors" and "unknown third parties" and follow them to a much greater extent than is the case with "information on websites of companies and brands" and any kind of advertising. This makes it understandable that great importance is attached to the ZMOT when building up trust. What is the consequence? Companies are increasingly losing **control of the information about brands and demand**. The power over the information is increasingly at the hands of the customers. Due to the fact that they access this information from price comparison portals and consumer forums more and more from a mobile device—i.e., directly at the stationary POS—clever shopping using the ZMOT is not restricted to online users.

> ▶ **Remember Box** The **ZMOT** has become a key **success factor** during the **customer journey**. At the same time, the ZMOT can purport either a recommendation or even a rejection of a concrete offer. We as a company must actively take part in creating this ZMOT and must not underestimate it any longer.

Two examples illustrate the relevance of the ZMOT. If one had searched for an *A3 convertible* with an IP address in New York in the summer of 2014, *Google* would have presented the corporate website of a car showroom in first place of the organic list of hits. On the one hand, it becomes clear that this company has conducted an excellent search engine optimization to be positioned right at the top of an organic search. At the same time, it could reveal that the management have probably only attended to the rating of their own performances by customers a little. How may a potential customer react when confronted with such a result during his search? This is where those in the companies concerned, who are responsible for this, are called upon to motivate their—hopefully existent—satisfied customers to post positive feedback. Doing nothing would be wanton.

This ZMOT forces companies to dedicate themselves to their own processes—and not only with a focus on the sales and, where necessary, post-sales areas, in order to reach a holistic positive **customer experience**. If this does not succeed, other potential buyers will be informed in the form of the ZMOT whether a company likes it or not! The company cannot stop this type of communication; at the most, it can try to become involved in it in the corresponding media. Sophisticated **web monitoring** is a mandatory requirement for this, in order to at least capture the relevant contents of this ZMOT communication at an early stage and be able to influence it where appropriate.

> **Think Box**
> - Of what significance is the ZMOT in the case of our business model?
> - How much information do our potential customers and customers retrieve about my company, our brands, and our offers on the Internet?
> - Which platforms do they use to do this?
> - What information do searching find there?
> - What tone towards our performances is dominating?
> - What indications are there for optimizing our performance?
> - Where can the responsibility for ZMOT management be anchored in my company?

Figure 2.10 shows the **intensity of the networking between offline and online channels**. 91 % of the people visiting a stationary shop buy there. But 65 %

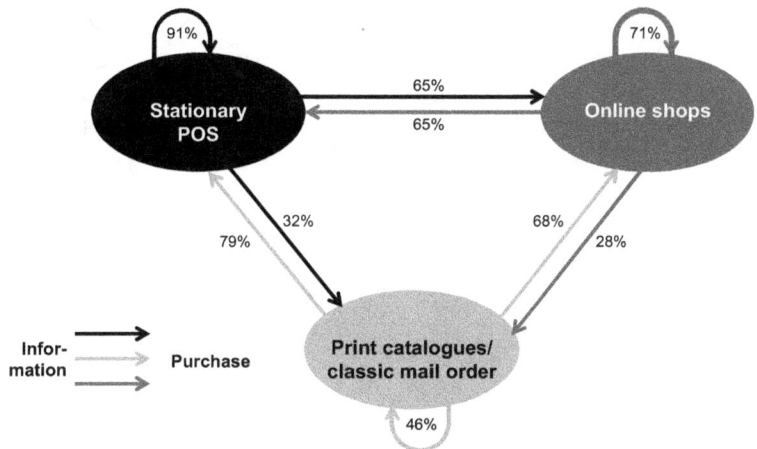

Fig. 2.10 Relevant information sources and their impact on purchases within the customer journey—usage in % (multiple answers possible) (*Source* Author illustration, data source: Kersch, 2013, p. 11)

purchase at online shops or use traditional print catalogs for placing their orders, too. The figure also shows that visitors of online shops buy offline (65 %) or based on printed catalogs (28 %). With this in mind if nothing else *Zappo* and many other digital pure players started up the large-volume dispatch of catalogs.

With that, a detailed evaluation of the customer journey in 2012 revealed that the majority of possible sources and channels of information give birth to the most diversified **customer journeys** (cf. Fig. 2.11). We thus need to find out what type of customer journeys dominate among our customers, in order to support them as much as possible with information and, where necessary, direct the resources towards the most important **customer touch points**. At the same time, it must always be taken into consideration that customers are increasingly becoming **multichannel customers**.

The big question is: Will we be able to sit in the "driver's seat" in the future when creating customer relationships and, in particular, when forming the customer journey? Or are the relationships reversing, and CRM becomes CMR—**customer managed relationship**?

> ▶ Remember Box **The customer defines which channels he uses in the course of his own individual customer journey**. Therefore the customer alone decides where he gets information from and where he makes his purchases in the end. If a company is not represented there, they will lose their position in the relevant set.

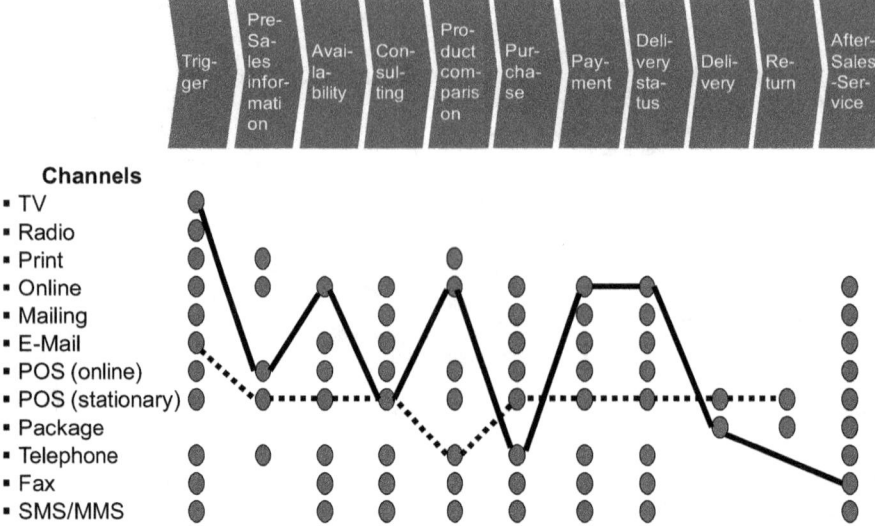

Fig. 2.11 Two customer-specific customer journeys regarding non-digital products (*Source* Author illustration)

Think Box
- Do we know the typical customer journey of our customers?
- Which channels and which sources of information are used particularly intensively—or are hardly made use of?
- Do we really know the preferences of our customers along the customer journey?
- Or do we only know how intensively the channels that we offer are used?
- Are we prepared for the customer deciding the way we have to walk when it comes to the information and distribution channels?
- Who in my company is responsible for Customer Journey Management?

With this experience in mind, the AIDA formula presented above needs to be consistently developed further in order to accommodate the additional activities during a customer journey. The result of this is the formula presented in Fig. 2.12: **ASIDAS**. The acquisition of attention for a certain offer is now frequently followed by an extensive **search**—which can lead to ZMOT. However, ASIDAS no longer represents an unflexible sequence of steps. In fact, the search process penetrates the steps of establishing "**interest**" and "**desire**" as well as triggering "action." In all steps of the process, feedback can be exchanged with friends and unknown third parties. "**Share**" happens parallel to or at the end of a customer journey. With this, all activities are described which cover the sharing of one's own experiences. This

Fig. 2.12 ASIDAS—the advanced AIDA formula (*Source* Author illustration)

can be via posts on *Facebook, Twitter,* or *WhatsApp*; by comments in forums, blogs (e.g., *Tumblr*), and communities; and of course in personal dialog.

Think Box
- Which platforms and which instruments do we offer for the potential customers and customers to go through the process called ASIDAS in such a way that they purchase from us and also give positive feedback about us?
- How extensive is our knowledge about what happens in each individual phase?
- In which sectors is my company active and how?
- Where can we arouse the attention of potential customers and customers?
- Where do they search, what increases their attention, where do they buy, where do they share their experiences and feedback? Enter your level of knowledge into the following form (cf. Fig. 2.13)!

Given all the euphoria about the involvement in social media, we need to visualize the **1:9:90 rule** (cf. Fig. 2.14). Studies show that—across all countries—approx. 1 % of Internet users are very active and post own contributions in blogs or online communities, for example, 9 % of Internet users respond to such entries—whereas a "silent majority" of 90 % is merely active by reading (cf. Petouhoff, 2011, p. 231). This means that we need to recognize the 1 % of **opinion leaders** on the Internet, in particular, and ideally win them over, so that the ZMOT works for us and our offer.

What can we derive from this information? The decisive point is that we attach much more significance to an extensive **customer journey analysis** than has previously been the case in many companies. In the process, our primary task as a manager is to identify the customer journeys preferred by the customers—even if they relatively clearly deviate from those processes we have planned. A great challenge for many companies comprises first and foremost defining effective **customer touch points** and to recognize their (positive or negative) contribution within the scope of the customer journeys (cf. Fig. 2.15). In general, touch points are those points that exist between stakeholders (i.e., potential customers, customers, employees, suppliers, cooperation partners) and our company. Figure 1.11 demonstrates the diversity that needs to be considered these days.

Actions by	Attention	Search	Interest	Desire	Action	Share
Company						
Customer/ Prospect						
Company						
Customer/ Prospect						
Company						
Customer/ Prospect						
Company						

Fig. 2.13 ASIDAS—appraisal for your company (*Source* Author illustration)

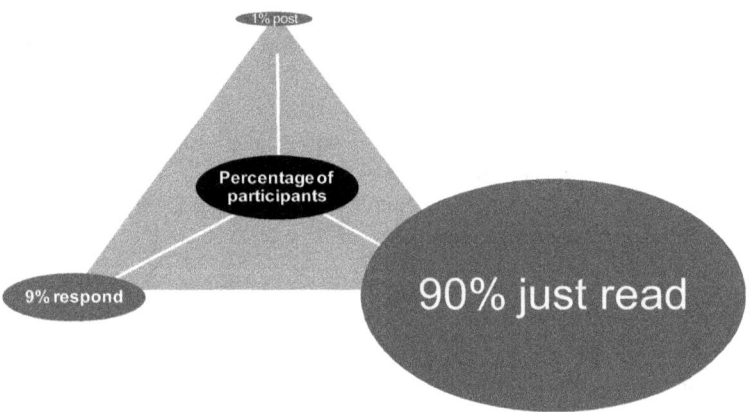

Fig. 2.14 The 1:9:90 rule (*Source* Author illustration, data source: Petouhoff, 2011, p. 231)

There are two types of customer touch points, by means of which a potential customer or customer can come into contact with a company, to be differentiated. These are, on the one hand, the **customer touch points of the company's own sphere**. These include the classical communication channels such as TV, radio, billboards, newspapers, and magazines. Spots, billboards, and adverts can be designed both monologically and to prompt a dialog. In addition to that, the following dialog media belong to this type of touch points: phone calls, mailings, emails, the corporate website, apps, online advertising, brochures, customer magazines, and the POS. Apart from that, the customer service center as well as corporate involvement in social media also belong to the company internal touch points. Yet also—merely seemingly—profane things represent touch points: In the case of e-commerce, these are the parcels delivered and in the case of many

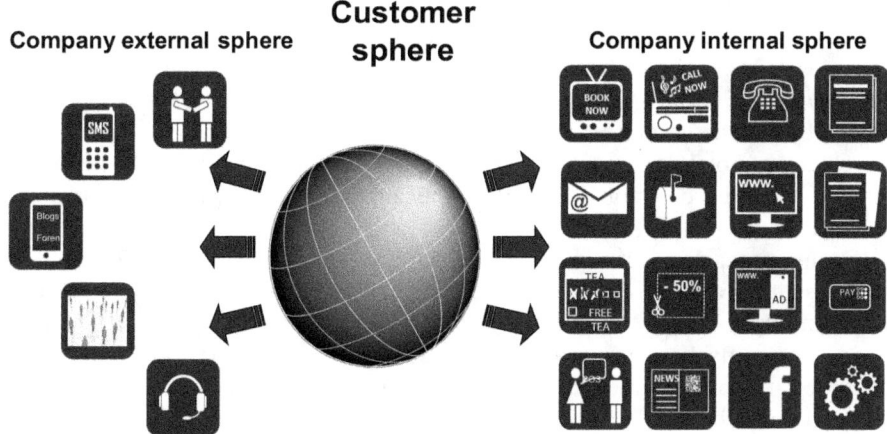

Fig. 2.15 Advanced concept of customer touch points (*Source* Author illustration)

providers, the packaging of the product itself (*Apple* and *Montblanc* are masters of this). But also invoices—frequently sent in a rather cold manner by mail or email—are important touch points.

On the other hand, there are the **customer touch points of the company external sphere**, on which the companies only have limited influence. This includes the communication within one's circle of friends, which can be activated company—and brand-specific by a tell-a-friend concept. It also covers the involvement of customers in blogs, forums, and social networks. As a company, we can and should try to have some influence here; it is however limited.

In the scope of corresponding analyses, we have ascertained time and again that the managers in the company were **not aware** of many of these **customer touch points**. As well as that, the **number of** relevant **touch points**—from a customer's perspective—was regularly **underestimated** considerably. This experience is confirmed by a study by *Esch*, for which 106 marketing decision-makers were asked. About half of the respondents presumed less than 50 touch points. In contrast, the study identified more than 100 touch points for most companies (cf. Esch, 2012, p. 3). And—of similar importance—it is important to know the perceived touch points from the customers' point of view. A recent study—based on 1,250 interviews—showed that the POS, the completion of a contract/process of payment, the homepage, personal advice, and a written offer are the touch points with the highest awareness level (cf. Keylens, 2013, p. 6). Yet how should **holistic customer relations** be achieved if not even the number—not to mention the contents and awareness—of these touch points is known?

In the course of own analyses, it has been regularly revealed that many touch points—and not only those previously unknown—are not managed sufficiently. This is regularly the case with the touch points of the ZMOT in particular. In addition to that, the significance of individual touch points is frequently evaluated completely differently from a company's and customer's perspective. This is where we need to determine—across the entire customer journey—how important the various touch points actually are for the potential customers and customers in the pre-sales, sales, and post-sales phase. According to a study by the *Internationale Hochschule Bad Honnef*, 59 % of all users allow themselves to be influenced "very much" by Internet criticisms and a further 37 % "somewhat," when, for example, the choice of hotel and traveling offers is involved (cf. Ludowig & Schlautmann, 2012, p. 7). It becomes obvious: the significance of the ZMOT can no longer be denied.

Furthermore, the following must be pointed out. The touch points with the strongest effect after the product are the direct—frequently also personal—**contacts with employees from our company** (cf. Esch, 2012, p. 6; Keylens, 2013, p. 8). These frequently have the most long-term effect—positive as well as negative. However, it also holds true that the employees actually in contact with the customer are frequently the most poorly paid. And these employees have the least information about significant developments in the company and the core values of the brands. It's worth thinking about this!

> **Remember Box** Without extensive stock-taking of a company's own touch points and their effects on the customers, no successful **customer touch point management** can be conducted.

Think Box
- Have we ever done extensive research as to which touch points exist in my company and our brands?
- Of what significance are these various touch points from a customer perspective?
- Which touch points are relevant in the pre-sales, sales, and post-sales phase—and for which groups of customers?
- How extensively are the particularly important touch points managed by my company—from a customer perspective?
- What measures are suitable for the optimization of touch point management?
- Which KPIs can be used to determine the success at the various touch points?
- Who should be responsible for this customer touch point management?

In order to fulfill this task, the topic of **Expectation Management** needs to be shed light on. Here, we also speak of **customer experience management**. We need to realize that through our communication (in particular in advertising), we constantly establish expectant attitudes with the addressees. Incidentally, this holds true not only in business life but also to the same extent in private life. Anybody promising "delivery within 48 h" and takes 4 days to deliver knowingly produces

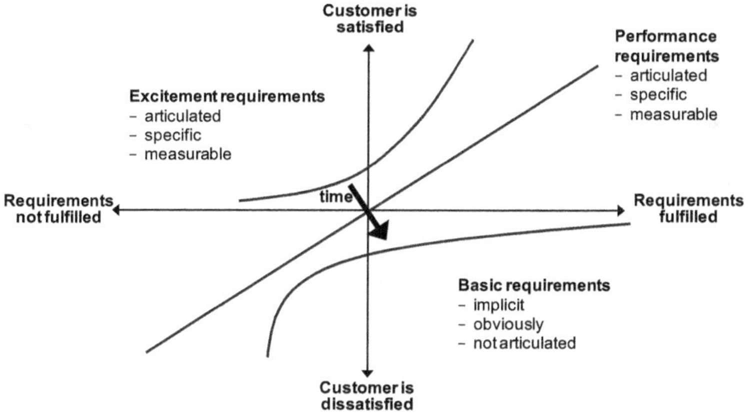

Fig. 2.16 Kano model of customer satisfaction (*Source* Author illustration, data source: Berger et al., 1993, p. 26)

unfulfilled expectations. For this reason, it is our quintessential responsibility to guide the expectations of customers into an area we can satisfy. Only those who accomplish more than promised will trigger excitement as illustrated in the following.

The so-called Kano model illustrates the significance that can be attached to the various services of a company for the so vital achievement of customer satisfaction needed for long-term company success (cf. Fig. 2.16). *Kano* examines the relationship between the fulfillment of customer requirements and the achievement of customer satisfaction (cf. Berger et al., 1993). In the course of this, it became clear that a proportion of the customer requirements have no or little impact on customer satisfaction (cf. the curve at the bottom in Fig. 2.16). The non-fulfillment of such requirements called **basic requirements** does indeed lead to dissatisfaction; the fulfillment of these does not lead to satisfaction or excitement. The fulfillment of them is taken for granted without making a contribution to satisfaction in the long run. In an online store, for example, this includes functional payment systems. Customers thus take the fulfillment of these basic requirements for granted.

Performance requirements are evaluated by the customer according to the principle "the more, the better." An excess of performance requirements increases satisfaction (cf. the middle line in Fig. 2.16). In the online store, this includes, for example, the number of products offered and the possibility to be able to save various selected products. It is not until the third category in the shape of **excitement requirements** that the excitement of the customer can be triggered, because then, services are rendered, which were not expected. If such services are frequently rendered, the risk is, however, that these mutate into performance requirements and are then expected (cf. the upper curve in Fig. 2.16). In an online store, free delivery for the first order, a small gift in the parcel or the indication of price reductions for the products marked as interesting, can trigger excitement. Frequently, though, it is usually only a question of time until such services become performance requirements.

> **Think Box**
> - Is it known in my company—from a customer perspective—what basic, performance, and excitement requirements are, respectively?
> - If this knowledge is still lacking, how can we close this information gap as fast as possible?
> - How do our competitors succeed in gearing their services to the various requirements?
> - Who can be entrusted with the implementation of the Kano model at my company?

The **triggers of an involvement in social media** from a user perspective have already been presented with the terms entertainment, education/growth, save money, save time, and solve something. These triggers come along with a phenomenon which can justifiably so be called a **social revolution**. It can be described as:

▶ Public is the new private!

More and more aspects, which in the past rather remained hidden, are being dragged into the public eye. A lot of impressive examples could be found on *webfail.at*. And this trend towards "public" and thus "social" is continuing. In the course of research work for this book, we intensively analyzed the relevant specialist and public media and identified a downright **social hype**. A flood of new terms with the word "social" put in front of them became visible. While discussing "social media" for a long time, in the past few months, new coinages have constantly cropped up. Figure 2.17 provides an overview of the most important terms with the word "social" put in front. The term **social landscape** is used for this because the landscape in which companies are active nowadays has shifted towards "social" in the long run.

The drivers behind many developments are applications called **social software** (also cf. Li & Bernoff, 2008a, b). At the same time, it holds true that "The social software of 2.0 is an attack on the established rules of power and forces fundamental rethinking" (Stüber, 2010). The most important—and long known—characteristics of this social landscape are the **social networks**. With *Facebook, Twitter, Google+,* and *LinkedIn,* they are enjoying increasingly more attention with regard to time and content—by users and companies equally.

Major significance is also being attached to the concept of **social sharing**. Thanks to concepts such as *YouTube, Vimeo, Flickr, Snapchat, Pinterest,* and *Slideshare,* it is becoming possible to upload all kinds of video and image files and make them "socially" accessible. Social software also facilitates the development of **social news sites** such as *reddit* and *NewsVine*. These allow registered users

Social Networks

Social Software

Social POS

Social Badges

Social Bookmarks

Social Log-ins

Social Events

SOCIAL COMMERCE

Social TV

Social Sharing

Social Recruiting

Social Filter

Social CRM

Social Ads

Social News Sites

Social Intranet

Social Search

Social Pressure

Fig. 2.17 Social landscape—the inflation of the word "social" (*Source* Author illustration)

to post links or own contributions which others, in turn, can rate. A special concept for recommending online contents is represented by **social bookmarks**. These involve online users placing bookmarks for websites that are interesting and which it should be possible to access in the future without conducting a new search. As these bookmarks are managed by social bookmarking services such as *Delicious, Mister Wong, StumbleUpon, and digg*, they are visible to the outside and thus "social" because they are shared with friends and like-minded people. The so-called social badges make a significant contribution to the previously mentioned social media. These are requests on the Internet, placed by users and companies, to turn potential customers and customers into fans, followers, or "pinners."

New developments, in contrast, are platforms such as *Eventful, Meetup,* and *Upcoming*, which provide support when organizing **social events. Social commerce** is enjoying increasing importance. A suitable definition of social commerce is "Help people to connect where they buy and to buy where they connect" (Marsden, 2011). Friends are thus connected virtually when shopping (e.g., via *Facebook*). Here, so-called *Facebook* mirrors are being tested—like in clothes stores such as at *adidas*. Photos of people in the selected outfit can be taken with a digital camera at the POS in order to present them right away to friends via *Facebook* or *Twitter*. And instant feedback can be reckoned with in this always-on generation. And while still in the store, the virtual thumbs of friends can go up or down and thus influence the purchasing process (cf. Geisler, 2012). In this way, the POS is becoming a **social POS**—even more so than before. This can also be successful that the number of *Facebook* Likes—here in a flagship store of *C&A* in Sao Paulo—is displayed in the clothes hangers, in order to be of "social" help when purchasing.

> **Think Box**
> - How can we involve social networks even better to create added value for our customers and added value for my company?
> - How can we advertise for my company on the social sharing platforms with our contents?
> - Does the use of social bookmarks create value for our customers and thereby for us?
> - How can we place social badges?
> - How can we create our on- and offline POS in a "social" way?

What significance social networks already have as "currency" today is shown in the concept of the social log-in. With **social log-ins**—also called social sign-ins—users use log-in information already available from social networks such as *Facebook* and *Twitter* to log in to third-party websites. By means of this, the user does not need to set up a new account for every new platform protected by a log-in. This has a decisive advantage for the user: **convenience thanks to a one click log-in**. For the providers—both on the side of the social networks and the third-party websites—this form of log-in comes along with exciting additional information. For nothing other is being achieved than a **one-to-one tracking of the ongoing activities of online users**. This possible use is frequently permitted by clicking the Terms of Business—not read by the majority of users—as read and accepted. And with each new access, further exciting information is gathered which, all in all, provides further fuel for social commerce.

Even the search processes on the Internet are becoming increasingly "social." Even today, with each of these search processes, there is a **social search**. The evaluation and consideration, controlled by algorithms, of searches by others form the basis for the autocomplete function—for example, on Google. By means of this, the passive search engine turns into an **active indication engine**, as in this way it becomes transparent to the (non-biased) user what other users have frequently searched for in particular. One thing becomes clear here: the suggested search terms can influence the user's further surfing behavior in the long run. And it also holds true that if, for example, customers search for negative reports about companies and, in the process, enter terms such as "deceit" or "mafia," they are automatically added to the search terms of the companies concerned. This also happens even if there is nothing to the allegations, because *Google* merely (predominantly) displays the search interest without evaluating the contents. However, *Google* and other search engines influence how reality is perceived in the long run!

The process of the **social search** also describes an online search where further user data are taken into consideration during the search process. The classical algorithms of the search engines determine the relevance of web content merely in view of the defined search terms. In the case of the social search, the relevance of web content is also measured according to whether it is of interest for the specific user. **Social metadata** are evaluated to this end. These are, for example, like statements by the user, social bookmarks placed and/or commented on, or other labeled or tagged contents. Thus the users' preferences influence the results of the organic list of hits of the search engines.

In the further course of the social revolution, the relevance of the **social filter** also increases. The consequence of such a filter, on the one hand, is that I am only shown what my friends also like. On the other hand, the social filter also has an effect via the previously discussed phenomenon of the ZMOT. The perception of the potential customers and customers is thus characterized more and more by the experiences of the social environment. This means that companies are increasingly **losing the control over the perception of brands and offers**. This now increasingly lies in the hands of the potential customers and customers themselves (cf. also the explanations on the filter bubble in Chap. 8).

It is interesting that with the increase in the relevance of the social filter, the **private filter** seems to be becoming less important. Thoughts, feelings, impressions, and evaluations are being sent by individuals—often spontaneously and without reflection—into the digital airwaves. When there they are visible for everybody for a digital eternity—with no sell-by date. Many millions have already regretted their **digital spontaneity of expression** in retrospect, because partnerships, friendships, business relationships, and work relations were put to an end after such posts!

Closely connected to the social filter are the so-called social ads on *Facebook*. The outstanding feature of these adverts is that they take the social context into consideration in which recipients of advertising can be found. Thus, social ads use the social influence of private networks as well as the **opinion leader & follower structure** found within them. At the same time, the various types of social ads are differentiated. One type could say "*Karl-Heinz* and three more of your friends like *Heinz Ketchup*." Another variation involves "sponsored stories" whereby the posts by companies are sent out to the circle of friends by one of them outing himself as a fan of the famous watch manufacturer *Lange & Söhne*, for example.

The **social CRM** is of particularly great importance. Here, **social media services** and **social media technologies** are used to intensify the interaction between customers and companies even more. The core of it is the intensification of the interchange between a company and its customers, which goes far beyond the primary relationship in dialog form of a classical CRM and thus involves instruments, channels, and contents from the social online involvement of the customers. This topic is dealt with in more detail in Chap. 7.

An exciting development for "social" is also revealed with TV. Under **social TV**, we understand the connection between the classical TV and social media (cf. Chap. 3 for more detail).

> **Think Box**
> - Do we want to use social log-ins to learn more about our customers?
> - What consequences does a social search have for my company? And how can we benefit from that?
> - What could we achieve via social ads?
> - How can we benefit from the trend towards social TV?
> - Who analyzes the implications of these developments on my company?

In the case of **social recruiting**, the social networks are involved in the search for new employees. The initiative can come from either the employee or the employer. Particularly widespread is the job search via *LinkedIn*, *Facebook* (with *BranchOut*), *Google+*, and *Twitter*. Whereby with this, a new—and frequently also more affordable—way of accessing certain target groups exists, one danger must not be ignored. Former and current employees can—partly protected by anonymity on the Internet—exchange information about the employer and, in doing so, spread internal matters that are not suitable for the aspired **employer branding**. At the same time, however, employers have the chance to see how applicants present themselves in blogs, online communities, or in social media. In the USA in 2014, it could be observed that employers asked for the password for the access to *Facebook* during the interview, in order to analyze the applicant's environment. This led, in our opinion rightly, to an uproar in the media, because in this case privacy was not being observed.

The increasing networking between people leads to a massive increase in **social pressure**, too. Where failures and mistakes made by companies in the past were only registered by a few insiders, a corporate failure today, can be distributed across the world in only a few minutes—by (former) employees, customers, competitors, bloggers, or journalists. The wave caused by a viral distribution of real or supposed "sins" of companies emerging is called a **shitstorm**. If people are severely criticized online, we speak of **cyber mobbing**. Instead of that, our ambition should rather be to initiate a viral **rose shower**—as an alternative draft to a shitstorm—by involving social pressure.

Even the intranet is becoming "social" so that one increasingly speaks of a **social intranet**. This means the internal exchange of knowledge within the company which makes use of social media. By means of this, the exchange of knowledge between the company's employees is simplified and accelerated—as the exchange of information via emails with long lists of recipients and even longer replies to the many questions posed by others is avoided. It is thus possible to dynamize static content by integrating in social media and to accelerate the social interaction. Employees working together on projects can get together via closed user groups within social networks. The exchange of information can be organized at the same time via internal wikis, blogs, forums, and/or *Facebook* groups—even beyond country and time borders. Newsfeeds keep those involved in a project up-to-date and make sure that all of those involved are at the same level of information at a certain point in time. At the same time, these mechanics promote a high degree of involvement—for even the respective superiors can obtain an overview of the status of the project's progress at all times. Ideally, a better informed and more strongly focused team is reached by means of a social intranet. This is particularly successful if the access to these social platforms can be made irrespective of time and place. Corresponding tools are offered by *jivesoftware.com*, for example.

> **Think Box**
> - Which possibilities of social recruiting do we employ or should we use?
> - How can we prepare ourselves for social pressure?
> - What potential does social software offer for the setup of a social intranet in my company?
> - Who could check the relevance of these developments for my company?

The **social revolution** has only just started, yet the upheavals caused by it are already becoming visible. As companies, we are well advised to contemplate the challenges associated with it at an early stage. Even here, it holds true to do the first finger exercises, so that we are prepared to be pulled onto the **social field**. For that is where everybody is watching whether one falls flat on their back or leaves the field as a winner.

> **Think Box**
> - Of what significance are the different forms of the social revolution for our business model? How can we benefit from it?
> - Where are we at risk of being run over by the social revolution?
> - Who bears the responsibility of evaluating the challenges for my company described by the social revolution?

▶ **Food for Thought** **For a start, all customers are in search of good feelings!** Good customer support, a wide range of goods, convincing advice, or a successful appearance in social media can contribute towards building up these "good feelings" just as well as handling personal data in a responsible manner can. For in the end, it is all about the positive emotional connection between your company and your customers. No more, no less!

In view of the **challenge posed by the digital and social revolution**, there is one thing we must bear in mind: The **selection process of digital Darwinism** starts when systems and processes in the economic system and society change faster than companies are able to adapt. At the same time, it holds true that Every company, every brand, and every offer is vulnerable. No business is "too big to fail" or "too small to succeed" (Solis, 2012b). With that said, the following aspects apply:

- The knowledge available in recent years is depreciating enormously. This also means that the **success stories and best cases of the past** can no longer be sustained in the future.
- The **currency of experience** is being systematically inflated and thus depreciated by new developments. This is why many companies are revealing enormous resistance towards the upcoming changes. For it means leaving appreciated zones of comfort.
- In many fields there—still—are **no comprehensive methods of measurement and metrics** to make it possible to measure the economic results. Yet, this must not lead to being not responsive to new challenges.
- **Marketing** will have to change dramatically in order to fulfill its role **as a strategy and energy driver** in the future.

> ▶ **Food for Thought** Many the one person or other person concerned should remember this guiding principle:
> **"In danger and dire distress, the middle of the road leads to death."**
> *Friedrich Freiherr von Logau*
> *Poet of the baroque period*

The impact of the changements in consumers' and customers' behavior has a direct impact on the communication process. *Dirk Ziems* presents interesting insight of cross-media communication campaigns.

Article by Guest Author *Dirk Ziems*
On the Mechanics of Cross-Media Communication Campaigns

The new **era of digital, interactive, and social media** has initiated a movement from classical mass media brand campaigns to new forms of interactive cross-media brand campaigns. Empirical research on the impact and mechanics of the new campaign formats shows the rich creative potential of the new genre as well as the rules of this new chapter of advertising culture.

Cross-media campaigns have become standard in recent years. Whereas communications planning in the 1980s and 1990s often still focused on one lead medium—either print or TV commercials—it has been standard practice to design **advertising campaigns as integrated communication campaigns** since the 2000s.

Based on numerous qualitative psychological market research studies on the subject, the mechanics of cross-media campaign forms will be described from the perspective of **media psychology**. The findings are based among other things on studies for *IP Deutschland* and *Seven One Media*, the two major TV advertising marketers in Germany.

Cross-media campaigns were traditionally a simple linking of the media channels by repeating or more closely specifying the announcement of a product or promotion in a commercial in print or Internet media. An example is that the *Mon Cheri* chocolate season is declared open by a TV commercial and the subject repeated in magazine print ads. For the case that all possible media channels are used for the same brand communication message (TV, radio, print, out of home, web), the term **360° communication** is commonly used.

For about 10 years, a new creativity in playing with the media channels has found its way into **brand communication**. In this connection, just think of the large number of so-called e-tales like the *Hornbach DIY* store campaign with the invented stuntman *Ron Hammer* from 2006 (cf. crossmedia.de/cases/hornbach-ronhammer) or the *Mark Ecko* fashion designer campaign with the faked story of the *Air Force One* graffiti sprayer likewise from 2006 (cf. wired.com/techbiz/media/news/2006/04/70718). Both examples make use of the same communication mechanics: generating news with something incredible that has never been seen before and exploiting this news for the purposes of the own brand. Strictly speaking these mechanics can already be traced back to one of the founding fathers of public relations, *Edward Bernays* (cf. Ewens, 1996, p. 15). In the case of *Hornbach,* the incredible is the motorbike stunt on the occasion of opening a new DIY store that leaves the biker unscathed despite explosion and long fall. In the case of *Mark Ecko,* it is the apparent ease with which the masked man can approach the *Air Force One* machine of the president and spray it with graffiti. As a result, the video hoaxes become a talking point, quickly spread via *YouTube* and social media, and are also discussed in TV news and talk shows. Until in a second phase the hoax is resolved, the brand (or the creative adman assigned by the brand) appears on the scene and transfers the hero or outlaw values to itself.

Cross-Media Story Telling
The e-tales or hoaxes reveal a **mechanism of digital cross-media communication** that points way beyond the classic linked campaigns. The story or advertising idea is no longer just told and repeated in different channels, corresponding to the classic 360° communication scheme. Rather, in the **digital and interactionist cross-media campaigns**, the advertising idea develops cumulatively like in a chain reaction through interaction of the various media, formats, and channels. In media sciences these relations are also termed "**transmedia story telling**" (cf. Jenkins, 2006, p. 97). Therefore, successful cross-media campaigns are characterized by a high level of efficiency. The spark of the *Hornbach* campaign was ignited, for example, by just a few airings. The *Air Force One* video of *Mark Ecko* was only placed in *YouTube* and immediately began to lead a life of its own with discussions in news broadcasts, interviews with security staff, and general speculations in the media. The campaigns quickly connect all possible channels: TV commercial and social media, PR and sponsoring, dialog and POS (cf. Mahrdt, 2009, p. 41f., 117f.). Whether the spark of **e-tales** and **hoax campaigns** catches or not is hard to predict. It is thus understandable that cross-media campaign types like co-creation and casting are now used far more often than the electronic tales.

Backgrounds on the Trend Towards Cross-Media Campaigns
Before making a more detailed typology of cross-media campaigns, a closer look is to be taken at central backgrounds first of all. The trend towards cross-media campaigns is to be understood as the result of crossing two main lines of development. Firstly, the **development from advertising recipient to advertising user**: consumers deal in a more relaxed and confident way with advertising nowadays. The times when the consumer was critical of advertising and resisted the deluge for fear of being manipulated and influenced by advertising seem to be a thing of past.

Today, consumers basically see themselves in a strong position. Kreutzer and Land describe the new **emancipation of consumers** from the supply and competence function of market providers in this book with the formula "**Me, Everything, Immediately, Everywhere.**" "Me" means that the consumer takes it for granted that he is the center of attention. He naturally expects appreciation and personalized offerings. "Everything" and "immediately" mean that there is nothing in a long-tail economy that can't be gotten immediately. And "everywhere" means that consumption is not fixed to place and channel and becomes ubiquitous within the digital possibilities.

Given this feeling of emancipation, **advertising communication** is just another of the many thematic and communication offers which is taken up autonomously and received as **part of wider entertainment and development offers**, shared with others (sharing via social media) and played with interactively.

These relations become particularly clear when recalling the new **viral potential of communication**. Every ad that has high talk value has a high viral potential, too. **Viral commercials** have long held top positions in the hit parade of the mostly widely viewed commercials, e.g., the roller babies commercial of *Evian* which, with over 100 million clicks, is one of the most viewed full-video commercials (cf. digitaltrainingacademy.com/casestudies/2012/11/evians_roller_babies_the_viral.php).

Secondly, the **development from mass media paradigm to the interactionist media paradigm**: classic advertising is decisively shaped in its sociological and psychological functions by the classic mass media paradigm—more than the admen and marketers usually realize. According to this, there are the powerful manufacturers and brands that have sovereignty over the media channels, whereas the consumers are part of an anonymous crowd of recipients, without any feedback channel or interactivity worth mentioning. Being at the mercy of the mass media and constantly confronted with its advertising manipulations also triggers instinctive advertising criticism and defense as backlash.

By contrast, the relationships change fundamentally in the new digital media era shaped by the Internet: the **advertising recipient**, who is the receiver of messages and is more or less passively involved in, **becomes the advertising user**. The **new advertising paradigm** builds on **symbolic interactionism**—the campaign message is started up, developed, or even fleshed out in joint interaction between brand and brand community (cf. Fig. 2.18; cf. Blumer, 1969, p. 94).

> **The new interactive communication paradigm: recipient as collaborator in the advertising process**
>
> Impulses of the campaign interact in a chain reaction from advertising recipient to other advertising recipients.
>
> → **Viral effects**
>
> Within the chain reactions the boundaries between passive and active, medium and everyday, fiction and reality, play and consequence become blurred.
>
> → **Hybrid effects**
>
> At the same time the various media channels are linked up through interaction.
>
> → **Amplifying effects**
>
> As a consequence, the overall interaction of the consumer/user community results in a holistic message of the brand campaign, which creates a motivating and involving brand community spirit.
>
> → **Brand community effects**

Fig. 2.18 The new interactive communication paradigm (*Source* Illustration by Ziems)

An example for illustration: in a casting contest from 2011 *McDonald's* called on its customers to "Build Your Own Burger." **Gamified creative elements** on the microsite further encouraged and motivated the customers to join in the casting. For example, someone could put a photo of themselves in the back of a stretched limousine taking a test drive to the award ceremony of the casting winner. The burger casting subsequently went through a number of selection steps (customer jury, test kitchen, voting) until the five winner burgers were actually sold in the restaurant and advertised in all channels (city lights, TV, radio).

The most widely aired TV commercial of the campaign gives a particularly good insight into what part of the **cross-media campaign interactivity** it was important to "build" on: The viewer sees proud *Alex*, an average customer in a baggy sweatshirt who could have been overlooked in *McDonald's* just yesterday, and how he observes other customers enjoying his creation, the *"New McPretzel with Meat Loaf." Alex* goes up to the customers, pats them on the shoulder and asks "Do you like it?" (cf. youtube.com/watch?v = iaUo5F0nWUw)

What does this commercial want to tell us about the point of the campaign? *McDonald's* is not just concerned about delicious tasting new burgers but wants its customers to feel accepted and in good hands with their preferences and wishes. *McDonald's* offers its customers to even gain a **field for shared self-realization** (new extended emotional end benefit on the brand benefit ladder).

The campaign opens up a **playground of interactions** between the customers and the brand leading to the customers interacting with each other as a community. Instead of "customers recruit customers," the warm emotional message of "customers feed customers" arises. *McDonald's* becomes self-referential as a brand—a **feedback culture of brand ambassadors** among themselves.

The *"Day One"* campaign of *Prudential* from 2011 is another good example of successful **interactionist cross-media story telling**. The initial spark that *Prudential* ignited at the start of the campaign was to request its own pension product customers to take a photo of the first sunrise in their new life as pensioners and to reflect on "day one" of their new stage of life. *Prudential* was thus offering its customers room for reflection that otherwise often falls short in the ambivalent subject of being put out to pasture. The US American customers evidently enjoyed taking stock, telling stories about their life and their plans, and presenting their environment and favorite places. At any rate, thousands of videos and tons of photos accumulated on the online platform of the campaign. In a final commercial, all the photos were compiled into a joint collage (cf. youtube.com/watch?v=162Qv2O2LK0). It conjured up a common image that, in many variants, holds together the **spirit of the brand community**, i.e., the image of reframing the concept of old age. Getting old and retiring is not the dark tunnel in the *Prudential* community. The ideal verified by thousands of customers is that of the new beginning repeated a 1,000 times over, with more than 6,000 sunrises on average still awaiting the US American customers after reaching the pensionable age.

The Variety of New Possibilities
In the **new cross-media world of communication**, many different possibilities of campaign creation are presented because different interaction schemes converge with different modes of effect. The **cross-media campaign orchestra** is a vivid model illustrating these relations (cf. Fig. 2.19).

- **Classic TV commercials** are one pole in the field of possibilities. Here the advertising user remains in the classic and comparatively passive reception mode.
- **Viral campaigns** basically remain in the reception scheme—with the difference that campaigns in social media networks are further spread by the recipients themselves.
- **Interactive online ads** build on restricted interaction (cf. example *Tipp Ex Buzzmann*, wuv.de/kampagnen/kreation_des_tages/interaktive_spot_kunst_von_tipp_ex). Only few interactive ads have proven successful in practice.
- **Apps**: Prolonging moving image campaigns in smart phone apps and games appears to be more promising and now more up-to-date. One example is the *"dumb ways to die"* campaign of the *Metro Melbourne* (cf. youtube.com/watch?v=IJNR2EpS0jw). More than 70 million views demonstrate the impact.

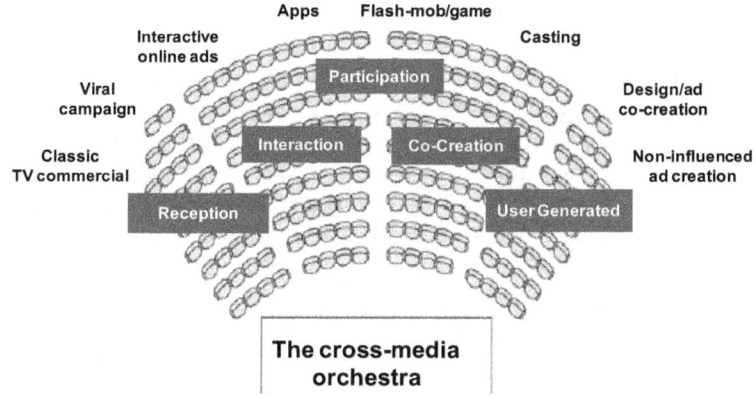

Fig. 2.19 The cross-media orchestra (*Source* Illustration by Ziems)

- **Flash mob and games**: The central mechanism of this campaign type is staging a participation of normal everyday consumers in a brand event. An example of this is the variety of English *T-Mobile* flash-mob campaigns based on spectacular choreographies in railway stations and airports (cf. one at *Heathrow airport* youtube.com/watch?v=ZMG2vNVq0ww).
- **Casting**: The named example from *McDonald's* shows the motivating power a casting offer has on consumers. Here the consumers/customers become direct collaborators. Another of numerous examples is *Seven Eleven Slurpee—Bring-Your-Own-Cup-Day* (cf. youtube.com/watch?v=OVXQniWfzig).
- **Design/ad cocreation**: Besides the cocreation of new products (see *McDonald's*: Make Your Own Burger) the cocreation of commercials is a motivating proposition. The ad is based on user-generated content. Aside from *Prudential*, think of the *Telekom Million Voices Project* where *Telekom* initiated the biggest choir of the Internet (cf. youtube.com/watch?v=cZIB6UK9OiE).
- **Non-influenced ad creation**: Forming an extreme pole in the orchestral fan are campaigns which develop on their own from the Internet community uninfluenced by companies or organizations. An example of this is the women against breast cancer campaign in which thousands of women show their breasts on the Internet (so-called mamming) to draw attention to the need for screening (cf. focus.de/digital/videos/virale-kampagne-gegen-brustkrebs-internet-trend-mamming-fordert-brueste-raus_vid_42150.html).

Control of Sovereignty Over the Brand

The **diagram of the cross-media fan** clearly reveals that the brand loses its classic authoritative role, the more consumers' participation and creation increases. Practice shows the need to observe a critical limit beyond which **companies lose control over the sovereignty of the brand**. This critical limit comes at the threshold from casting to the non-influenced user-generated content.

This relation is made evident by two examples:

- **Negative example** of *Pril* packaging casting: what turned out to be a flop of innovative cross-media campaigns was the washing-up liquid casting of Pril in 2011. The Internet community was asked to create their own original designs for *Pril* bottle labels. To the dismay of the marketing people, the community chose the design "chicken fragrance" showing an ugly chicken as the winner. Of course, *Henkel* could not put this product on the supermarket shelf and was hence forced to cancel the Internet casting.
- **Positive example** *McDonald's "Build Your Own Burger":* Here *McDonald's* subtly incorporated restraints in the casting process allowing the brand control over the chosen products. Particularly worthy of mention in this respect is the test kitchen run by *McDonald's* where burgers with too expensive ingredients were silently eliminated from the contest.

The Key Role of Full-Video Commercials in Cross-Media Campaigns
Practice has also shown that **cross-media campaigns** can only seldom manage without the medium of **moving images** and **TV as channel**. This applies for most, not all, cross-media campaigns. An exception from this rule is, for example, the brand *Red Bull*, which bases its marketing on event marketing causing a sensation as such. Background to the **dependence of most cross-media campaigns on classic commercials** is the particularly emotionalizing power of full-video commercials. Neither with static print visuals nor with online interactivity is it possible to reach the involvement in emotionally moving story telling that is possible with full-video commercials.

Emotional story telling is the basis for igniting the big idea that gets the cross-media campaign going and initiates the subsequent cross-media chain reactions. Therefore, full-video commercials often form the start to cross-media campaigns or play the role of summarizing the overall interpretation of the campaign (see above: *Prudential* campaign, *Hornbach* campaigns). However, the question is raised as to whether **full-video commercials** will necessarily be tied in the long run to **linear TV programs**. Full-video commercials can also be integrated in other media like *YouTube*, social media, etc. Movements away from linear TV are currently taking place among media users, the consequences of which cannot be fully foreseen.

Dirk Ziems, Managing Partner concept m research + consulting GmbH, www.conceptm.eu

Quick Wins

- _____
- _____
- _____
- _____
- _____
- _____
- _____
- _____
- _____
- _____
- _____
- _____
- _____
- _____
- _____
- _____
- _____
- _____
- _____
- _____
- _____
- _____
- _____
- _____
- _____
- _____
- _____
- _____

3 Big Data and Technology: Drivers of the Information Revolution on the Part of the Companies and Accelerators of the Era of Cooperation

Back in 1957, *Peter F. Drucker* spoke of the so-called infoworkers in his legendary book *Landmarks of Tomorrow: A Report on the New Post-Modern World*. This is what *Drucker* called those employees who generated their added value within the company solely with the help of information. But it is not until now that we have really arrived in the **information era**. That is why the term infoworker is experiencing a renaissance at the moment. A forecast for 2020 states that at that point in time, more than 85 % of the world's working population will be working as infoworkers (cf. Schmidt, Göbbel, & Bchara, 2012, p. 38). The change in the world of employment with a new form of **networking the employees** with each other but also the **networking between employees and customers as well as other service partners** is based on the availability of information at all times. And how does the corresponding saying go for the existing supply of information?

▶ Everywhere, any time, at low costs!

Nowadays, more than 850 million people all over the world already work almost exclusively from a mobile device (cf. Schmidt et al., 2012, p. 38)—from a smart phone, a tablet PC, or a laptop. This frequently happens irrespective of country borders and time restrictions. The classical borderlines of the circles of performance between providers as well as their suppliers and customers are becoming increasingly blurred. The **value added chains** are interlinking to become increasingly complex **value added systems** in order to produce more efficiently.

▶ And the currency is called information

At the same time—not only with us and our employees—an increased **blurring of private and business life** can be ascertained, because one increasingly does not succeed in "classically clocking off." For why do companies attempt to prescribe their employees mobile phone and email free evenings and undisturbed weekends?

Because everything is merging and we are becoming increasingly unable to escape from the continuous **information overload**. Have we already established the information infrastructure in our company to accommodate this challenge?

Or do the gloomy forecasts, according to which **the infoworkers' increasing appetite for information** is becoming a nightmare of any ERP system, rather apply to us? Yet, at the same time, the infoworkers spend 80 % of their time on research and preparing the data, because they "search but don't find" because the **infoworkers' search logic** is not compatible with the **provision of the data**, because far too much time is necessary for combing through databases, preparing the information, and consolidating the data, because the data is not mapped. Are our infoworkers also spending only 20 % of their valuable working hours on utilizing the data and ideally generating added value for the company and customers (cf. Schmidt et al., 2012, p. 38)? Not only the providers of hard- and software are confronted with this challenge but all companies where **information** is proving to be a **strategic resource** and a **key force for competitive advantages**. And the development towards big data will intensify this data boost substantially.

The challenge lies in the **setup of knowledge-based jobs**—also functioning in the virtual and mobile world—where, ideally, the infoworkers' need for information is anticipated and the development of tailor-made solutions and offers for customers is supported. A particular challenge comprises the overcoming of the **system and structural interruptions** that still frequently exist in companies, which exist due to varying hard- and software solutions. These obstacles can be overcome by hard work and money.

In contrast, the **obstacles in the heads of the managers and employees** are much graver—and at the same time significantly more difficult to diagnose and rectify. In many cases, they are downright **cognitive firewalls**, which stand in the path of cross-department, cross-area, and/or cross-border collaboration. This leads to the feared **silo mentality**, which often blocks the **view of the whole picture**—in particular when it comes to strategic issues. For top performers too frequently cling to their resort egoisms and tend to be more interested in optimizing their "own small-small"—always with the next round of bonuses in mind. And unfortunately, this silo mentality frequently begins at a board or executive level and is then happily adopted by the downstream management levels by means of observational learning and lived out to the full. With that, various **types of silos** must be differentiated:

- **Data Silos**

    In this case, databases existing in companies are not merged to ensure a "holistic view of the multidimensional customer." Due to the fact that customers can be active online and offline on different platforms, merging them would pose a great challenge. Thus, the data acquired online about customer preferences or billing data, for example, are to be interconnected with the CRM system just like the data from social networks. 69 the data from social networks.

- **Process silos**

    One can always speak of process silos when company-internal processes are not properly attuned to each other. This frequently relates to the areas of product development and marketing/sales. Likewise, there is frequently a lack of a process-related interface between the customer service center and product development in order to act on suggestions and criticism by customers. Yet even between marketing and sales, there are still gaps to be overcome.

- **Silos in the heads of people**

    The most serious silos, however, are the ones in the heads of the management and employees (the "cognitive firewall"), because they frequently represent the cause for the aforementioned data and process silos. Thus the question can be posed as to whether a company-internal cooperation is rather promoted or obstructed by the own company culture.

We should ask ourselves whether we can afford an **organizational structure with a silo mentality**—partially even cast in cement. For what are the consequences of such a silo mentality? On the one hand, **company-internal opportunities** remain unexploited. On the other hand, downright **data graveyards** emerge—protected, but not used!

In order to overcome these bottlenecks, information is again necessary in many cases—yet this time **information** that is initially **internally directed**, which should contribute towards not only the management but all employees being aware of the business objectives and the strategic approaches but also convey to them the necessity of extensive cooperation. In doing so, this should not only be implemented in the company but also involve all relevant stakeholders. Sometimes such a company conception is called **Enterprise 2.0**. Exciting solutions, which steps can help a company to change in a change management process, will be presented in Chap. 9 to master the digital transformation.

> **Think Box**
> - How do we master the merging of private and working life in my company—at the expense of or in favor of the employees?
> - How much time do our employees spend on looking for information—instead of adding value by utilizing it?
> - How well have we succeeded in setting up knowledge-based jobs?
> - How pronounced are the data and process silos in my company?
> - Who is defending them and who can break through them?
> - Who can be assigned with a critical evaluation and the development of solutions in my company?

So much for the internal side of the information revolution. Yet, the actual **information tsunami** is still in the offing! And it won't be triggered by us providers or the sales partners. It was set off by the customers and is already approaching us. The first signs can be seen on the horizon. And one thing is certain: this information tsunami will not only wash over a large number of existing business concepts, it will destroy them for good! For in many cases, there will be no long processes of adaptation. What is awaiting us can be called a **disruptive change**.

But what does this mean for our company? Are we in danger of going under in this tsunami or could **value added chains** be developed at an early stage **by involving customers and other partners**, in order to use the power of the tsunami to energize our own turbines? The requirement for this is that the turbines are of the right size and equipped with qualified personnel. In this way, consumers can progress to become **prosumers** to act as performance partners, who equally consume and produce and thus integrate their energy in the corporate value added chain (cf. Chaps. 4 and 7; cf. Solis, 2012a).

As already demonstrated in Chap. 1, there is a large number of developments that will lead to a continuous increase in the flow of data. This is why one can rightly speak of **big data**. The fascinating thing is that since the year 2000, it has become cheaper to store a data in digital form on hard disk than on a piece of paper! With that, the limits to costs, which until that time had restricted **access to an extensive digitalization of information**, were lifted. Yet what are the impacts of this on all of our lives?

Nowadays, every child is born into a **digital world**. Unlike 40 or 50 years ago, when digital data and information were still the exception, today the ultrasound images of pregnant women are posted on *Facebook* and thus made available to the entire world. Merely a few minutes after birth, the first photos—still in the labor ward—are posted. Furthermore, from the minute of becoming pregnant onwards, further data is generated by, for example, blood tests, etc. This means that the first **digital footprint** is already made prior to birth. And in the course of life, further digital data trails, which we create in the course of our education, during various sporting activities, regarding work, etc., are added. By being involved in social networks, we reveal our preferences as well as our professional and private networks. And when shopping—either online or offline—much more information about us is collected. Maybe we are checking in at various real places (such as at *Starbucks* or at an airport) or TV programs and, via a possible **want function** on *Facebook*, are documenting everything we would like to have.

Such data, which is mainly available in digital form nowadays, can be processed to provide exciting findings—sometimes with more, sometimes with less effort. As a private individual, it can scare you—as a company, we frequently wish to have access to at least a part of this **gigantic digital data flow**—also called **big data**.

If one tries to obtain a deeper understanding of the term **big data**, it must first be pointed out that in this case, we are speaking about **large amounts of data** which cannot be processed by classical databases and data management tools or only to an inadequate extent. The great challenge comprises collecting the various different **data formats, update rhythms,** and **data sources** in order to transfer them into a relevant data flow. This challenge can be described by the following **dimensions of big data**:

Volume (as in data volumes or amounts of data)
- "Volume" generally describes the gigantic **amount of data** acquired due to the various digital footprints of man and machine. On the one hand, it is about the amplitude of the available data, on the other, also about their profoundness.

Velocity (as in the speed of the change cycles)
- "Velocity" describes at what speed new data sets are created, existing data sets updated, or deleted altogether. The means of data acquisition, which are reflected in the necessary loading and update times, also have an impact on the change cycles. At the same time, systems in which the changes are collected and recorded in real time are gaining importance.

Variety (as in the diversity of the data sources and data formats)
- Ultimately, "variety" means, on the one hand, the multitude that exists in the internal and external **data sources**. There are—first of all—the company's own data pools with information about transactions, for example, collected via the likes of sensors. Public data pools (such as generally accessible data pools) could be exploited as well. And now social media creates a high additional information density. On the other hand, there is a variety of diverging **data formats** and nomenclatures regarding the termini used. This makes a very high-performance mapping necessary so that no mistakes regarding the content are made when merging data. In addition, consideration must be given to the different **data memory locations**, because besides storing data locally, increasingly larger amounts of data are stored in the cloud. Further, the variety of the **data storage systems** (such as in terms of data model, type of database, hardware) must be considered. After all, the data also reveal—depending on the storage locations, for example—diverging **data availabilities**.

Each of these dimensions of big data harbors a multitude of further challenges. In some documents big data is mistaken for "**deep data**" (as in deep data structures), such as can exist in ERP or CRM systems. However, in contrast to big data, deep data lacks sufficient amplitude regarding the integrated data sources (such as in a data warehouse approach) as well as the various data categories. For this reason, deep data itself cannot create a holistic picture—of a customer, for example—as in a **digital footprint**. It is the merging of the various data sources and data categories—to be protected by data policies—that create the much aspired digital footprint by means of a **cross-linked evaluation of the systems used**. To this end, the incoming data in the stationary store as well as in the online shop must be pooled, linked up via a customer card as the case may be. In addition, the **data trails of the customer journey** on the company website need to be evaluated—with mobile or stationary devices—and linked up to the data of the CRM system. All of these steps are to be conducted with the goal in mind so as not to cause data graveyards that remain unused and unattended. The **increasing number of data users** additionally intensifies the challenges of big data. These users are not only companies but also millions of consumers acting both as senders and recipients—and doing this on completely different channels.

Where does the **key use of big data** lie for our company? The focus can be placed on five **areas of benefit**:

- **Increased transparency through big data**

    A systematic analysis of big data targeted towards the requirements of the respective business model enables the **recognition of relevant market trends** at an early stage. This can succeed on sales and supply markets at the same time. By analyzing big data, companies can thus achieve significant **competitive advantages** if interesting sales potentials are recognized and addressed specifically—faster and more extensively than the competitors do. In addition, new fields of business are ideally identified faster than other companies do, if one succeeds in "distilling out" wishes and expectations from the data. *Amazon* is working on "anticipatory package shipping"—i.e., they try to forecast the next order of a customer and ship the corresponding product already in vicinity of the "forecasted customer" ready for delivery—to increase speed. *Apple* just applied for a patent to base the delivery of mobile ads on the mood of the receiver (cf. Campillo-Lundbeck, 2014, p. 17). And the US retailer *Target* is able to predict a pregnancy—before other family members are aware of it!

- **Cross-validation of the data**

    If a **cross-channel merging and evaluation of the data** is successful, the reliability of findings can be increased by comparing the results from various sources. By doing so, the prognostic relevance of the statements can be increased.

- **Testing the various marketing measures**

    The abundance of big data makes it possible to be able to **test marketing measures** more extensively than was the case in recent years. However, for this purpose, **more powerful analysis systems** are required to gain "important insights" from the diversity of information. This also includes, for example, continuing to bring risk management to perfection in the case of flows of payments and trade by means of a systematic monitoring system.

- **Personalization and individualization (in real time if necessary)**

    Information frequently provided in real time makes the relevant data available for an immediate **personalization and individualization of the communication** and, where applicable, also a **one-to-one service offer**. The extent to which this potential for individualizing services by **real time marketing** can actually be increased depends on the business models. These possibilities promote the development of marketing increasingly becoming a service and customers being "wheedled" by extensive information and offers. A holistic knowledge of customers and the market facilitates tailor-made offers the customers can no longer resist—and for which they may even be ready to pay more money (cf. Chap. 7 for more details).

- **Speeding up and improving processes**
  The information flows of big data can also improve **company-internal processes** in the long run. A necessary requirement for this is that these processes have been prepared for this information tsunami and companies are not steamrolled. If necessary, the **business models driven by information can be automated**.

Yet what exactly are the **data sources** behind these developments? Figure 3.1 shows a selection of them, whereby it needs to be mentioned that dozens of new sources are emerging every day and existing ones are disappearing. It thus becomes obvious that: **An important driver of big data** is the fact that people have a "social" disposition by nature. That is why they love announcing their opinions, wishes, hopes, and anxieties—and doing so more and more frequently in social media, too, and thus publicly. This is where what is generally known as **Zuckerberg's Law** (cf. Hansell, 2008) comes into play:

"I would expect that next year, people will share twice as much information as they share this year, and next year, they will be sharing twice as much as they did the year before," he said. "That means that people are using *Facebook*, and the applications and the ecosystem, more and more."

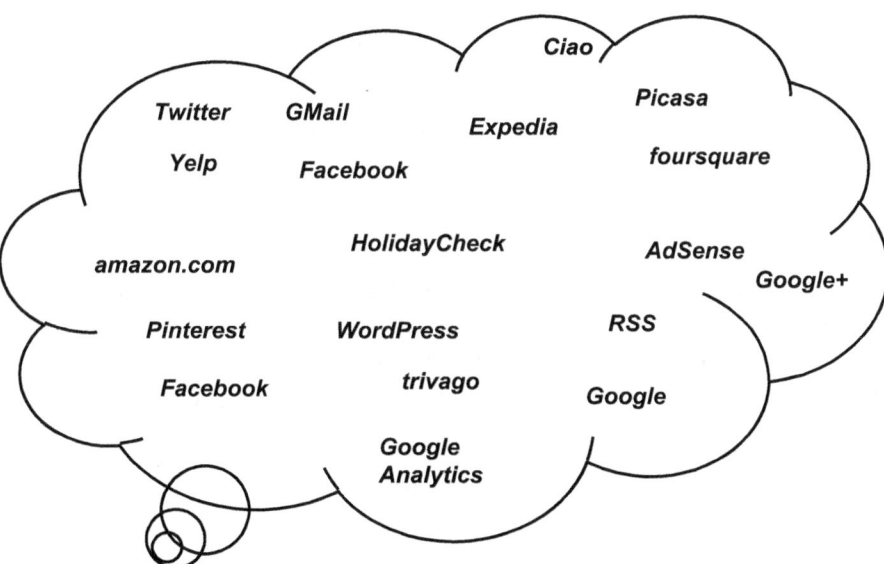

Fig. 3.1 Essential sources of big data (*Source* Author illustration)

The **voice of the customer** is being increasingly echoed in the cloud and can thus be analyzed more extensively and considered when approached and supported. The technologies required for this, such as "prescriptive analytics," "content analytics," and "predictive analytics," are available for precisely such applications and of increasing quality (cf. Gartner, 2013a) and will boost the relevance of using big data.

Latest research reveals which **number of cycles** there is in the various sources. Besides more than 340 blog posts per minutes, more than 72 h of video material are uploaded per minute on *YouTube*, almost 280,000 tweets sent, approx. 2 million searches started on *Google*, and about 2.5 million posts on *Facebook* are communicated per minute. Thus the database of *Facebook* grows by 350 gigabytes of new data per minute. At the same time—and this may surprise or reinforce some critics of emails—almost 204 million emails are sent (cf. Qmee, 2013). The **trend of this provision of information** is **spiraling** across all channels!

This flood of information is further fueled by the increasing **use of sensors** which diligently generate new data. We can already use the term **sensory economy**. Correspondingly, every smart phone user (qua telephony) as well as every stationary or mobile Internet user quasi-represents a **human sensor** that constantly generates new data. And the fascinating question for every company is:

> ▶ Which of the data available here contains valuable information, which is relevant for the further development of the company?

We should bear in mind that not only the "original data", that is consumed and saved, is concerned. Increasing amounts of **metadata**, which tell us something about the type of this consumption or about behavior in general (e.g. "Where is a telephone call being placed and for how long?"), are appearing on the scene, thus creating a new quality of information. This results in what is today called the **digital shadow**. Accordingly, the information about us is already much more comprehensive than the information we generate ourselves! And this type of information is continuing to increase.

Yet at the same time it must be taken into consideration that the **data is generated in various formats** and that highly efficient systems are required to evaluate it. The great challenge comprises turning the data acquired from online processes, CRM systems, controlling, email and telephone communication, as well as from the seemingly never-ending whirlwind in social media into information relevant to making decisions—and ideally in real time. This task will keep us busy in the years to come—just like the establishment of classic CRM systems has done in the past decades! For the task has remained—only it has become far more difficult:

> ▶ The creation of a holistic view of the data of the multidimensional cross-channel-active and not necessarily coherently behaving customer

What does this task mean in detail? The **holistic view of data** indicates that we merge the various data flows—ideally into one system—in order to focus on the customer as a whole. During our consultations, we constantly experience companies that fillet customers information-wise because they, on the one hand, see the online side of the customer and then the offline side but, in terms of data and IT, never obtain an overall picture. And they do that even though the customer has granted permission to use the data. The **multiple dimensionality** expresses that simple segmentation concepts no longer take effect and that a fan status at *Jaguar* can be far away from a possible decision to purchase. Furthermore, the **channel bandwidth** available to customers nowadays must be considered. Having just been on *Twitter*, a quick email is written, a comment is posted on some news on *Facebook* before looking for an online or offline POS. And all of these activities of our customers do not need to reveal **any consistency**. That means that the customer may and will contradict himself in his actions, something that should not be the case when we, as a company, approach our customers!

> ▶ Food for Thought **The customer has always been put first, but now he is also expecting the appropriate services** because he provides us with the information necessary for this in great amplitude, depth, and up-to-datedness.

At least some consolation can be found in coping with the handling of big data. The **costs for storing data** have fallen and will continue to fall dramatically (cf. Gantz & Reinsel, 2011, p. 4). At the same time, a moderate growth in the **investments in IT** must be assumed. This can only really mean one thing: there is an increasingly more powerful infrastructure available to generate insights relevant for decisions from the large amount of data. And there is an additional consequence: IT is becoming more and more a strategic resource—enabling the company to manage the **digital transformation** successfully.

> ▶ Remember Box **It's all about data!** And with that, big data is not to be perceived as a "stable something" or quasi as a "thing." **Big data** is in fact an extensive **data stream**, increasing by the day, by the hour, by the minute, which needs to be understood as a process capable of exceeding all available IT dimensions. This is because increasingly larger amounts of data in an increasing width and depth are being provided, these being in a multitude of formats and across various channels at increasingly faster speeds. The task for companies is to crystallize relevant information from this information tsunami and to preferably do this faster than the competitors.

Whereas the decision-based preparation of data by CRM systems has already demonstrated good progress in the companies in the past, dealing with the previously mentioned mere endless **whirlwind in social media** is a challenge which is still difficult for many companies to master. This is why the possibilities of **social media monitoring** are now being addressed. This represents a part of web monitoring and is also called web scouting or buzz tracking. Buzz stands for the buzzing and humming as well as for the babel to be monitored.

One way of accessing social media monitoring can be by means of **online trend monitoring**, which is, for example, supported by *Google Insights for Search*. The use of this free service makes it possible to determine the relevance of topics, persons, products, and companies in terms of region and time to find out what is currently being communicated more or less intensively. This can provide an initial and fast overview of emerging trends.

One speaks of **social media monitoring** when social media, i.e., forums and blogs, are the focus of the analysis. It can determine, for example, how intensively topics that are relevant for a company are discussed. These may be issues of data protection as well as possibilities for protecting one's own home against burglars, if a company offers corresponding solutions. It can thus be ascertained what issues in what fields are "hot" and reveal an advertising relevancy as the case may be. At the same time it is also investigated where the own brand name is being discussed to then join in the dialog. Social media monitoring should direct particular attention towards posts made on the likes of *Twitter*, *Pinterest*, *Google+*, or *Facebook*. The triggers of shitstorms are particularly common in such messages due to the viral potential of circulation.

All in all, it is essential for most companies to not only systematically comb social media but the entire Internet by means of **web monitoring** for posts relevant for the company. These can be opinions, trends, feedback on own or third-party offers, product and service feedback, but also impulses for innovations.

A first and free possibility of web monitoring available is the use of *Google Alerts*. After defining important search terms at *google.de/alerts, Google* automatically generates emails when posts with the defined search terms appear online in the "public part" of the Internet. This enables you to receive news from certain areas promptly, to monitor or determine competitors or business trends, whether posts about yourself, your offers, or your company appear. Even setting up RSS feeds or using *Google Reader* can contribute towards building up the necessary issue tracker to bring relevant content to the surface.

The great challenge comprises not only determining the **number of comments** but also their **tonality**. This is where so-called sentiments analyses come into use. Their task is to separate positive posts from negative ones. Ideally they should manage to do this also with those bearing an ambiguous statement. This is, for example, the case in the following statement: "That was a really GREAT service!?" Now is that a commendation or a criticism with an ironic undertone? When classifying such posts, many professional service providers depend on semiautomatic sentiments recognition. In plain English, this means that a human eye (in cases of doubt) does the classifying (cf. for this *Radian6, BuzzLogic, Nielsen, Visible*). The information gained is frequently classified into the categories "positive," "neutral," and "negative" and bolstered with examples in corresponding evaluation reports. You can check the possibilities by using free tools like *socialmention*, checking the findings e.g., for the CEO of the *Virgin Group, Richard Branson*, or for the *U.S. Bank*.

> **Remember Box** The great challenge when analyzing and evaluating mentions on the Internet and particularly in social media is differentiating between **fact, opinion, and populism**! At the same time, another crucial question to be asked is: **What is the sender's intention?**

It is essential for every company that not only a trainee sporadically googles the company or product names to find out what is being reported about the company, their products, or offers. This is where the installation of a continuous **Internet monitoring system** is necessary, especially if the company has reached a critical size. The information gathered in different ways forms the background for the design of the **social media engagement**. For social media is not a "by-the-way business"!

Think Box
- To what extent is web and social media monitoring used in my company?
- What monitoring targets were defined?
- Which tools are used for monitoring?
- How and for whom is the information gathered on the Internet prepared?
- What consequences resulted from the gathered data?
- What experience was gained?
- Who is responsible for monitoring?
- Are there sufficient human resources and money available?

Monitoring primarily involves listening when others are talking. Yet besides this monitoring, there is an even greater task for the companies in order to gain further **access to big data**, i.e., the challenge of motivating potential customers and customers to give companies the **token** to access their *Facebook* data. The token represents the permission to access the **users' digital memory** as an opt-in. Due to the fact that on *Facebook* an all-time data depth and width that is totally up-to-date is becoming accessible, many company processes can access it to submit more and more suitable offers (cf. Chap. 7 for more details).

Besides big data, the parallel **technological developments** need to be considered which, besides chances, can also involve risks for companies. An exciting change is revealed by the previously discussed trend towards **social TV**. With an average of

4 h and 24 min of TV consumption per day and US citizen, TV has sustained its position as a medium even in the era of the Internet—despite all prophecies of doom (cf. Nielsen, 2013, p. 4). In contrast, every Internet user of the USA only spends 38.80 h per month online (cf. ComScore, 2012, p. 5). With an average of 30 days a month, this would correspond to a daily usage of 1 h and about 17 min. This means the daily usage of the TV is more than three times higher than that of the Internet. And the average TV consumption equates to a daily use of the newspaper (only offline) of only 22 min as well as magazines (only offline) of only 16 min. Trend probably further downwards! At the same time, however, various aspects have to be considered. As shown in Fig. 3.2, it is particularly the older target groups that remain loyal to television.

Yet, this is where a change is beginning to show. Whereas in the past, TV consumption represented the main activity of the majority of watchers, today, a parallel use of various communication devices can be observed more and more frequently. While the TV is on, either a laptop, smart phone, and/or a tablet PC are used. With that, the TV is called the **first screen**, the stationary computer the second screen, and mobile terminals such as smart phones or tablet PCs third screens. Due to the fact that TV consumption is increasingly supplemented by the parallel use of smart phones and tablet PCs, mobile terminals are increasingly taking over the position as **second screen**. This enables a **second screen experience** as in a parallel usage of various devices.

As already addressed, more than one third of US tablet owners and almost one quarter of US smart phone users use such devices as second screen while watching TV for looking up the information they got from the TV program (cf. Winslow, 2012). Overall it is estimated that the second screen usage is even higher (cf. Bauder, 2012). This data result from a representative media study for the USA by *Nielsen* in 2012. The US media survey consists of a panel of online users with a representative sample of 1,998 adults. This process is called the **merger of**

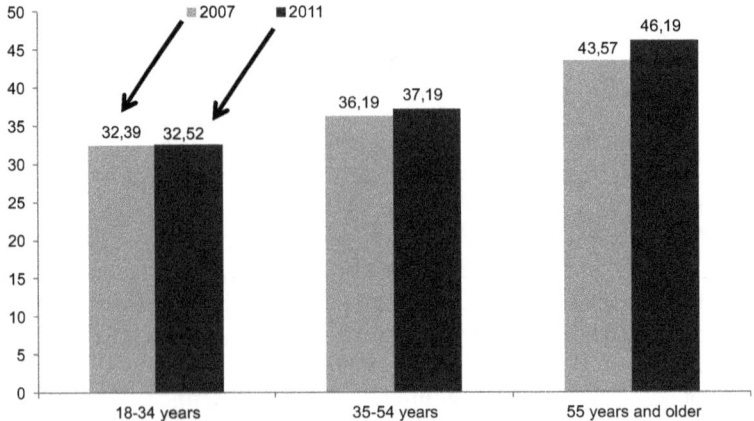

Fig. 3.2 Average weekly total use of television by US adults in 2007 and 2011, by age groups (in hours and minutes—HH:MM) (*Source* Author illustration, data source: Nielsen, 2012)

screens. On that note, it is conceivable as early as today that smart phones or tablet PCs—at least in parts of the population—are developing into the first screen.

With this communication running parallel to TV consumption, two developments can be differentiated. The parallel usage can be either synchronous or asynchronous to the TV. An **asynchronous usage of the second screen** is when searches are started on the Internet or emails are checked or the *Facebook* status is updated parallel to TV consumption but irrespective of the contents broadcasted. With **synchronous usage of the second screen** the user extends the contents presented on television by, for example, commenting the program being shown on the mobile screen via *Twitter* or *Facebook* and sharing one's own ideas with friends. Or one speculates together with friends about "who did it" on *CSI Miami*. The sound interface via *Shazam* makes it possible to link a TV commercial with additional information on the mobile device. A *Toyota* TV spot can hereby be linked to a test drive offer on the tablet PC or the smart phone.

According to the Nielsen study quoted earlier, in as early as June 2012, 33 % of US *Twitter* users exchanged opinions about the program being broadcasted. This corresponds to an increase of 27 % in only 5 months (cf. Bauder, 2012). In 2014 around 240 million people use *Twitter* at least on a monthly basis. During the TV duel between *Barack Obama* and his challenger *Mitt Romney*, 10.3 million tweets were posted on *Twitter* within 90 min, and the debate was extended consequently by the users in the social media (cf. Zschunke, 2012, p. 3). TV consumption can also directly prompt to buy awakening the desire of the music being heard or the dress being shown. Via a connection to social networks, advice can be obtained from friends as to whether the purchase should actually be made before buying the dress presented in a TV program.

Thus TV is becoming **social TV**. Applications such as *GetGlue* support such activities as they contribute towards supporting a personalized, networked, and thus social TV experience by means of a corresponding function on a smart phone or tablet PC by checking in when watching TV programs like it only used to be the case with physical locations (cf. GetGlue, 2012). The platforms characterchatter. usanetwork.com, zeebox.com, and wywy.com offer appropriate functionalities for the **combination of TV consumption and online communication**. These applications frequently alleviate access to certain programs, sports events, or protagonists—sort of like *Twitter*, where corresponding posts need to be searched by a hash tag (#). In order to make this easier on *Twitter*, some of the TV programs today fade in the corresponding # terms. The advantage with *Twitter* is that the users have access to their "fan community."

One thing becomes evident: it's all amounting to a **convergence of TV, smart phone, or tablet PC usage** and the **communication via social networks**, whereby *Facebook* and *Twitter* are especially integrated. The prognosis of the **Cluetrain Manifesto** (1999) is proving true:

> ▶ **Markets are conversations!** The clou in these conversations is to become integrated as an appreciative business partner with relevant

(promotional) information because, in this case, the person and context are known.

> **Think Box**
> - How can we shape this multiscreen experience such that it has an image and sales promoting influence on my company and our brand values?
> - What possibilities are there for my company to become involved in this ongoing communication with contents relevant for the users?
> - Who in my company can give the signal for this review?

In addition, a further change with TV is becoming apparent because television sets are being increasingly connected to the Internet (buzzword "smart TV"). In according to the report "Defining the In-Home CE and Network Ecosystem 2013" done by Diffusion Group, around 66 % of US households have Internet via broadband (cf. Mitchell, 2013). The percentage of US households having televisions connected with the Internet amounts to around 37 %. In detail, 25 % of US households connected to the Internet via broadband own a smart TV in 2012, which is an increase by 100 % within just one year. Due to the increased usage of mobile and stationary terminals, there is a further increase in accessing TV programs from the Net. A previously quoted study by Nielsen provides data about this. The cross-platform report shows the monthly time spent for watching TV of US inhabitants (aged 2+) for different media in 2012 in comparison to 2011. There is an increase of watching timeshifted TV in all TV homes by 6 %, and the time watching video on the Internet is risen by around 37 %. Mobile subscribers watching TV on their mobile phone is increased by 25 %, and there is only a 1.5 % increase by using the Internet on a computer (cf. Nielsen, 2013, p. 6).

The **smartization**, i.e. the "making intelligent" of devices via networking over the Internet, is becoming apparent: telephones turning into smart phones, TV devices into smart TVs, and watches into smart watches. This development will be fostered by additional so-called wearables—like *Google* glasses. And now smart refrigerators and smart washing machines are turning up, which start to work when electricity is at its cheapest. A particularly "smart" refrigerator can monitor its fill level in addition and automatically trigger order processes via the Internet—oriented to the preferred consumer behavior of the members of the household. There are even concepts for a smart house which can be controlled via the Internet from anywhere around the world. This entire development is also described by the already presented term **Internet of things** (cf. Chui, Löffler, & Roberts, 2010 for more details).

The distinctive feature of the **smart TV** lies in the intelligent combination of **entertainment electronics** on the one hand and interaction-oriented **possibilities of telecommunication and Internet** on the other. This integration enables you, for example, to look for and find a dress, seen and tagged in a film, in online shops and present it to friends via *Facebook*. There it can be voted upon whether it should be bought. The extension of TV consumption into social networks can lead to winning back the partially TV abstinent younger target groups of today. In addition, companies are confronted with the exciting possibility of presenting brands the minute they are being talked about via an evaluation of the ongoing communication performed in real time. This makes a **social online brand experience** possible that had until now been unthinkable. For only now are not only the users' preferences visible but also their specific moods (based on the TV programs watched). The days of **one-way TV** are—at least in younger target groups—counted. The challenge comprises first of all recognizing the social potential of **two-way** or **multichannel TV** as well as the trend towards **multiscreen usage** (i.e., TV and laptop, smart phone, and/or tablet PC in addition). Then it is necessary for the companies to integrate the new application possibilities in the ever more complex customer journey in an effective way, thus increasing sales.

The developments presented here demonstrate at the same time that business models that used to function side by side without any difficulty are increasingly competing with each other. *YouTube* is increasingly becoming a competitor of classic TV especially due to the recent start-up of special interest channels. (TV-) hardware providers are becoming content providers. The TV program providers will—supported by the social networks—join the domains of publishers even more and cannibalize there. In addition, the competition between existing rivals will become more acute. A classic mail-order firm that does not link their online offers with social TV in an elegant way will fall even further behind innovative online shops.

The challenge here is **seamless integration**. This means the "seamless integration" of various applications, the use of which so far could only be achieved by overcoming various complex interfaces. Ideally, an **ecosystem** emerges at the same time. This is a closed system that the user does not need to leave even if he wishes to start various applications. Thus, *Apple* offers direct user benefits with *iTunes* (with integrated cloud applications) as well as the use of new *Apple* products without any effort in integrating. The continued diversity of the applications of *Google*, *Facebook,* and *eBay/PayPal* represents further examples of this evolving ecosystem. For customers, these ecosystems offer a decisive advantage: **convenience**. And for the companies providing them, these systems involve two decisive advantages. A high level of **customer retention** develops, as the ecosystem drastically increases the **barriers for switching**. By doing this, high **barriers for alternative providers to enter the market** are set up at the same time.

> **Think Box**
> - What relevance can social TV develop for my company?
> - Where do I see interesting starting points to integrate it into my communication?

(continued)

- Where are there possible risks to be avoided?
- Do the ecosystems arising offer opportunities or risks if any for my company?
- Can own ecosystems be set up or is it more a matter of integrating in arising ecosystems?
- Who is assuming responsibility for closely examining these processes with their implications for my company?

Yet, which industries will be particularly affected by these developments "all about big data"? And how can these challenges be faced? Figure 3.3 shows the results of an **impact analysis of big data**. To this end, the various sectors in the USA were analyzed by *McKinsey* and positioned in a matrix. The **productivity growth** is shown using the USA as an example for the overall period of 2000–2008. The **big data value potential index** is defined by five criteria which reveal how strong an effect the dynamic sampling of the data stream has on the company (cf. McKinsey, 2011, p. 123f.).

Cluster A in Fig. 3.3 covers the areas of **computer and electronic products** and **information**, which have already greatly benefited from the development of "big data" and will continue to do so in the future. The two sectors in **Cluster B**—**finance and insurance** and **government**—can strongly benefit from big data if barriers of use are overcome. The sectors in **Cluster C**—including **construction**, **educational services**, and **arts/entertainment**—reveal a negative productivity growth. This can indicate barriers to a gain in efficiency. Interesting progress in productivity by big data can be expected in **Cluster D** in particular in the sector of **wholesale trade** and in the area of **transportation and warehouses. Cluster E** shows that in the sectors of **healthcare providers, real estate and rental,** as well as partially in **retail trade**, productivity growth can be anticipated thanks to big data (cf. McKinsey, 2011, p. 9).

How easy or difficult companies will find it to **lift productivity potential** is shown in the heat map in Fig. 3.4. The "overall ease of capture index" summarizes whether capturing it is rather easy or difficult. The results of the individual evaluation for the criteria "Talent" as in the "extent of qualified personnel," "IT intensity," "data-driven mind-set," and "data availability" have been incorporated. Here it becomes obvious that sectors such as **manufacturing, information**, and **utilities** will find it easier to reach the next productivity level. Sectors such as **arts/entertainment/recreation, government,** as well as from the field of **educational services** in contrast find it considerably more difficult to capture productivity potentials (cf. McKinsey, 2011, p. 10, 124f.).

The complexity in the company environment described here and also in the other chapters has another dramatic consequence: The **compulsion towards increasingly comprehensive cooperations** for the digital media have stretched the information density that everybody is assailed by every day to the limits of the tolerable. Never before have there been so many and extremely easily accessible possibilities

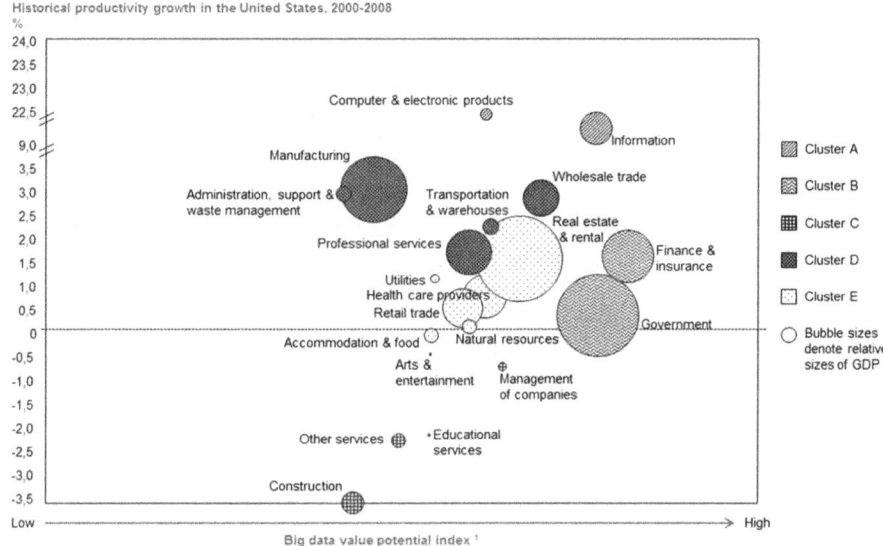

Fig. 3.3 Analysis of the impact of big data on different sectors in the USA (*Source* Author illustration, data source: McKinsey, 2011, p. 9)

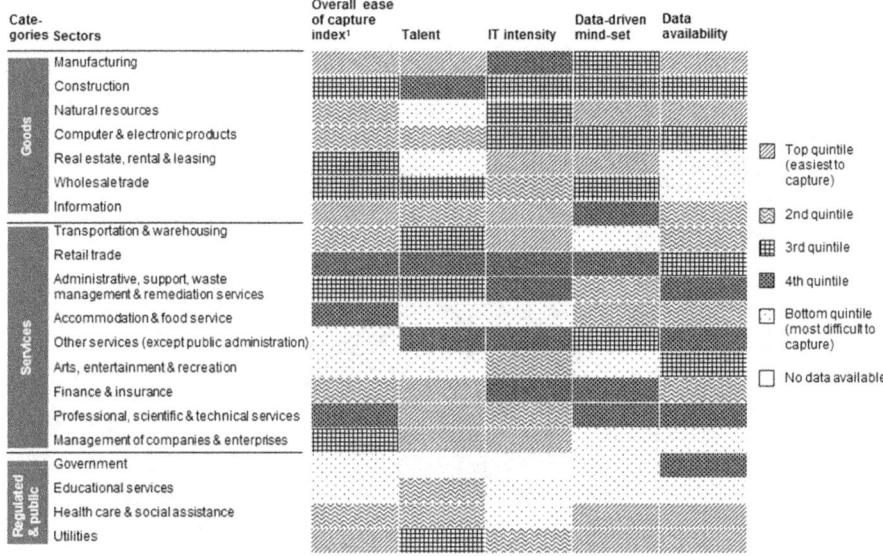

Fig. 3.4 Heat map—showing the relative ease of capturing the value potential across sectors (*Source* Author illustration, data source: McKinsey, 2011, p. 10)

to get information, entertainment, and to communicate (worldwide)—and when wishing to do so, all at the same time.

> Yet, if everybody can obtain information of their choice at all times, if everybody is taught to fight for their own small advantage in the economy, a system of innumerable individualists who indulge in their own interests will emerge. That forms such a complex society, the challenges of which, in turn, can only be tackled together. Thus part of the irony of this story is that the era of the individualists of all things conjures up the cooperation. Nobody else relies so much on the cooperation with others than the individualist. Society is the network of his life. He can only act out his individualism if functional communities assure him (Prange, 2012, p. 53).

The **necessity of cooperation** targets on the one hand the (former) competitors but also the customers and last but not least the interior of a company itself (cf. Fig. 3.5). The **involvement of customers** no longer takes place merely via concepts such as classic customer surveys or customer advisory boards involved in the organization, but much more extensively—and integrated in the values added process—via social media (cf. Chap. 4 for more details).

In parallel to that, the trend towards **cooperating with competitors**—even within the own sector or even within the own strategic group—must be recognized. The strategic group is formed by the companies of one industry, which have a comparable business concept with similar product ranges with similar price ranges

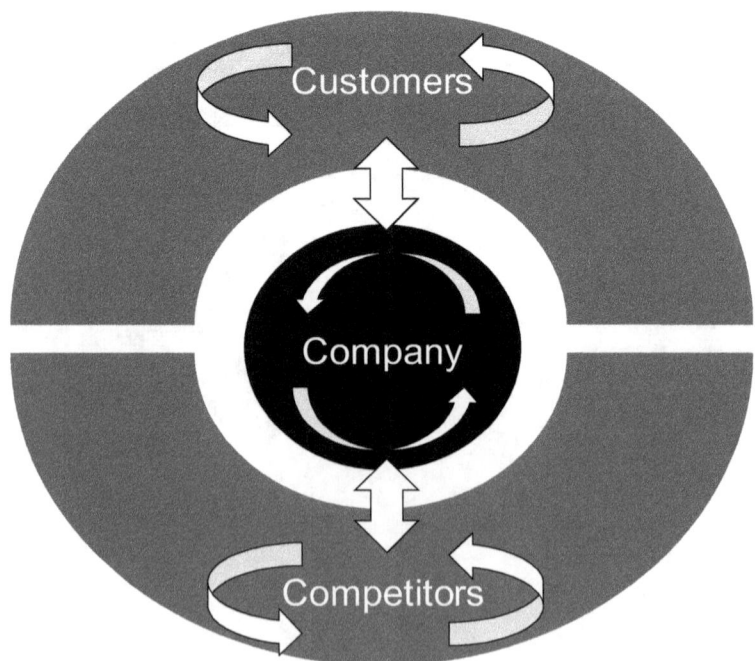

Fig. 3.5 The era of cooperation (*Source* Author illustration)

and address the same target group via related communication and distribution channels. Thus the description of the world by *Thomas Hobbes* as in a "war against all" or "dog eat dog" needs to be overcome.

For this reason, the Darwinian formula for success of "survival of the fittest," which is solely based on adaptability or strength, no longer applies. With **digital Darwinism**, it's all about **smartness** and **cleverness**, in order to be able to successfully survive in the increasingly complex and dynamic environment. The challenge here is to master the process of the **digital transformation**.

In the **cooperation between companies**, smartness is revealed by the cooperation actually being created to the advantage of both parties—even if the strategic competitive advantages are being set aside at first glance. The basis for such cooperation is **trust** (cf. Chap. 6). The reason for this can be provided by the **game theory**. It tries to elaborate under which conditions the players position themselves at the target of a larger benefit to be reached mutually. Usually, each player will tend to only play a small part if there is the danger of being ripped off by other players. It is not until trust in the other players also sticking to certain rules has been built up that the players will become more involved and open up more so that a significantly better result can be achieved. This cannot be managed without building up trust!

If trust can be successfully established, then even **cooperations between archrivals** are possible such as between *Daimler* and *Renault*, *BMW* and *Toyota*, and *General Motors* and *PSA Peugeot Citroën*. But there are even extensive cooperations between *Apple* and *Samsung* and *Boehringer Ingelheim* and *Eli Lilly*. Even *Facebook* grew from a variety of cooperations because it opened up its platform for developers, who wrote thousands of applications for *Facebook* at an early stage and who thus contributed to the popularity of the network. Sometimes such a temporary marriage ends again as can be observed between *Apple* and *Google* regarding *Google Maps* (cf. Hofmann, Fasse, & Postinett, 2012, p. 54). Even many conquests described in Chap. 1 would not have been possible without cooperations. This also goes, for example, for the entry of *Google* in hardware market with the *Nexus* series. This would hardly have been possible without the strategic cooperation with *Samsung* and *LG*.

The necessity of cooperation does not remain restricted to the sphere outside a company. In order to overcome the silo mentality described and the resort egoisms, the **corporate cooperation potentials** must also be recognized and exploited. The economist *Richard Sennet* very aptly put it: "Bonus systems are the enemy of any cooperation." An example of this is provided by the so-called friendliness calendar of investment bankers: "In March, very friendly, in July somewhat cool, September aggressive, December fight for yourself" (Prange, 2012, p. 53). This statement applies if at least one's own bonuses are only to be achieved in competition with your own colleagues and not with them. For this reason the question should be raised as to what extent the bonus systems established in the company are suitable for supporting cooperations—even beyond board and hierarchy levels (cf. Chap. 9 for details on the digital transformation).

Think Box
- How great is the openness in my company for cooperations with presumed archrivals?
- Can exciting cooperation fields be found here, which should be illuminated for once without blinkers?
- How open are we for cooperation with our customers? Are proposals for cooperation made really meant seriously?
- How well do cooperations work in my company—even beyond board and department boundaries?
- Do our bonus systems spur on corporate competition and thus cement resort egoisms?
- Or do our bonus systems motivate to look beyond the rim of the teacup and to try to not only discuss value-adding concepts with those responsible but to also put them to practice?
- Who is responsible for focusing on these various cooperation fields—internally and externally?

▶ **Food for Thought** Information is cheap! Sense is valuable! Why not recognize the meaningful and apply it as a basis for your own business strategy?

Quick Wins

-
-
-
-
-
-
-
-
-
-
-
-
-
-
-
-
-
-
-
-
-
-
-
-
-
-
-
-
-

How the Social Revolution Is to Be Managed 4

We are in the middle of the **social revolution**! By **using social media,** on the one hand, **social relationships develop between the users,** who encounter each other on the same hierarchy level. On the other hand, **opinion leader-opinion follower relationships** are developed, which become established by the mutual creation, further development, and distribution of contents via social networks as well as via blogs and communities. The low barriers of entry when using social media—such as low costs, easy possibilities of uploading contents, and simple handling (high usability)—promote the sharing of them. At the same time, the classic opinion leaders (such as journalists and analysts) become less important even if their own loss in importance has still not become evident to them and they regard a commitment to social media as being partially less relevant (cf. Wüst, 2013, for current study results).

The most important **utilization categories and sample applications of social media** can be found in Fig. 4.1. One group is formed by possibilities primarily targeted at **communication** such as blogs (like *Tumblr*), microblogs (e.g., *Twitter*), private and professional social networks (such as *LinkedIn, Facebook, Google+, Classmates.com*), social bookmarking platforms (e.g., *mister-wong.com, dig.com, stumpleupon.com*), as well as forums and communities. The focus of another group is on the **cooperation between users**. In this case, wikis are set up together (such as *Wikipedia* or *Wikileaks*) and existing services are evaluated in the scope of review and information portals (e.g., *ciao.com*) or newly created in the scope of creative portals. The third group covers **content sharing**, i.e., sharing content about specific platforms such as *YouTube, Snapchat, Flickr,* or *Pinterest*. These contents can be the likes of texts, videos, photos, or audio files. Such content sharing, however, also takes place in social networks because here, too, various contents are shared with others.

By linking the concepts demonstrated, complex **social media applications** can emerge, representing attractive communication platforms for companies. The important aspect is that all of these applications offer one thing:

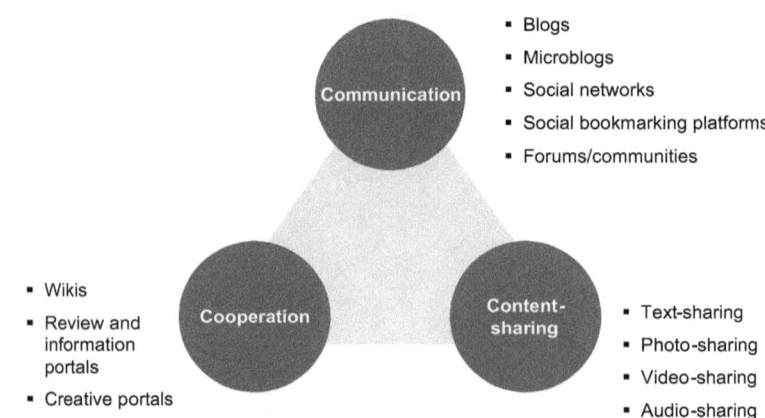

Fig. 4.1 Utilization categories and sample applications of social media (*Source* Author illustration)

The possibility emerges of a dialog between companies and their relevant target groups—at significantly lower costs and more target-oriented than ever before.

Social media thus offer the possibility of overcoming the classic company perspective from the inside to the outside ("**inside out**") and to at least supplement or overcome the company perspective by integrating external impulses ("**outside in**"). By doing so, the possibility of implementing postulated listening is dramatically extended by a variety of new channels. At the same time, we can overcome the classic innovation approach by grasping impulses from the market and society extensively and at an early stage and integrating them in our innovation processes.

Due to the increasing significance of the possibilities being offered here, the term **social media marketing** was introduced. Social media marketing describes the procedural concept that avails itself of the users' engagement in social media to reach marketing objectives. In doing so, it must be taken into consideration that the needs and motives described in Chap. 2 are availed of. Within social media, three **media categories** can be differentiated (cf. Mayer-Uellner, 2010, p. 16; Oetting, 2010, for the basics). The online activities the companies are responsible for are called **owned media**. These include the corporate website, email communication, as well as an online shop. The possibilities for communicating with the users via *Twitter, Facebook, Pinterest,* as well as corporate blogs and own forums and communities are also part of it (cf. Fig. 4.2). But we have to consider that the *Facebook* page of a company is not really owned by the company itself; it is owned by *Facebook*. Owned media must be managed in a target-oriented way (buzzword: **manage**), in order to develop them as communication channels that are at least to some extent able to compensate the loss in importance of TV and other advertising channels. The area of **paid media,** which describes the measures companies buy from third-party partners, must be distinguished from it. Examples of this are banners, sponsored links, sponsored stories, sponsored posts, promoted tweets, as

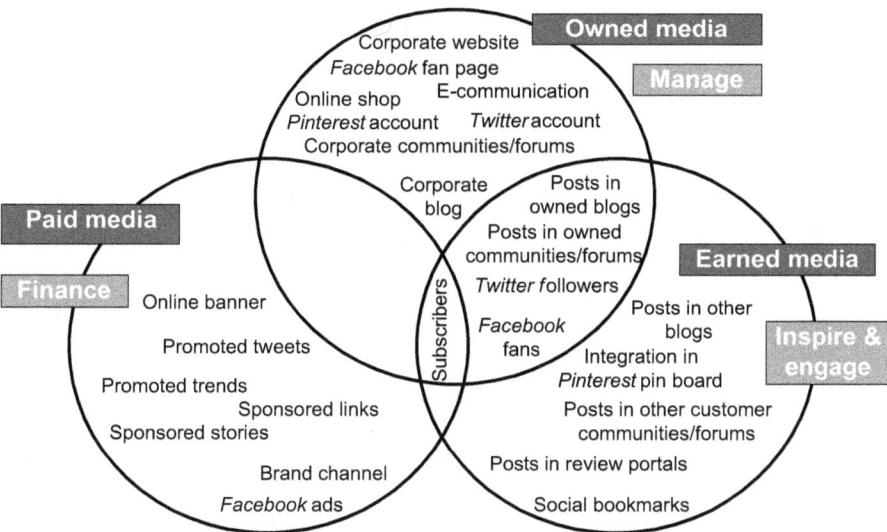

Fig. 4.2 Overall view of the different categories of social media (*Source* Author illustration)

well as channels in branding one's own company or own brand (e.g., on *YouTube*). Access to these possibilities is merely a question of finances invested (buzzword: **finance**).

The third category—**earned media**—describes the platforms as well as in particular the contents the companies have "earned" from the Internet users due to their activities, for better or worse. This is user-generated content in various shapes. It includes, for example, social bookmarks as well as posts in noncorporate customer blogs, forums, and communities as well as in social networks. An important requirement in order to achieve a high (positive) share in the area of earned media is to earn attention as well as engagement in social networks. Besides the necessary investment of time and money, the ability to tell good stories is also a part of this. This development is described by the term story telling or narrative marketing. The buzzwords in this case are thus **inspire** and **engage**. And one thing is sure: the relevance of earned media for companies will continue to increase—as here, authentic dialogs and "real" information are more likely to be expected because the sender of this information is not a company but "regular" users.

Many other contents lie in the **overlapping areas of the media categories**. If users are requested via join-in campaigns to create their own content on platforms operated by companies, this part of user-generated content belongs in the overlapping area between owned and earned media. The same applies if companies invite users to a dialog in a corporate blog or own forums and communities, and this is complied with. Even the fans gained on the likes of *Facebook*, *Google+*, or *Pinterest* and the followers on *Twitter* belong there. If users subscribe to a brand-specific *YouTube c*hannel, they belong as subscribers to the overlapping area between paid and earned media.

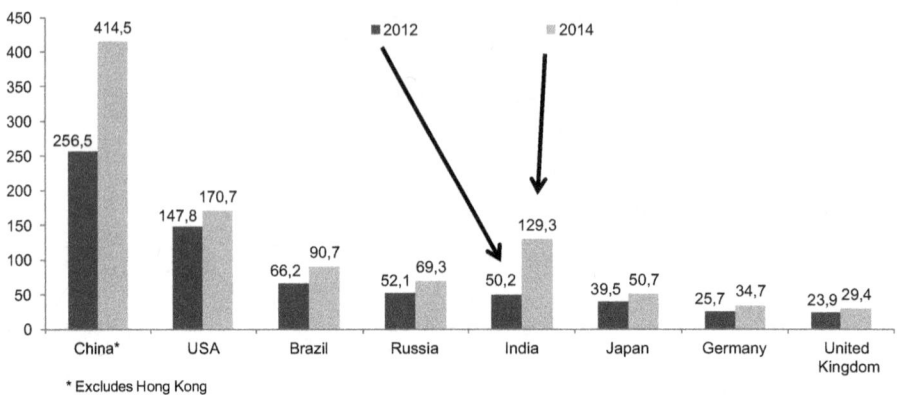

Fig. 4.3 Number of users (in millions) of social networks in selected countries in 2011 and forecast for 2014 (*Source* Author illustration, data source: eMarketer, 2012)

The classification illustrates initial important **differences between social media and classic mass media.** Whereas using classic mass media is reserved for professional users, engaging in social media is **open to every Internet user**. Another distinguishing feature between social and mass media consists of social media enabling **real-time communication** in many cases—both in terms of provision and the changing of contents. With that, a disproportionately higher speed in exchanging information is facilitated as is possible in most mass media due to the widely linear communication (cf. Fig. 1.13).

Figure 4.3 shows what **increase in importance social media** have experienced since 2011 and will experience in the future. It is expected that in the USA, 170.7 million users will be active in social media in 2014. As can be seen, dramatic growth rates are also expected in the aspiring developing countries such as China and India.

In order to find out the **significance of social media in the USA**, it helps to take a look at various studies regarding the user behavior of US citizens:

- In 2012, the USA had the **world's largest share of users of social networks** in comparison to the overall population (49.9 %; cf. eMarketer, 2012).
- In 2012, 22 % of US population (age 12+) used social networks several times per day, 15 % nearly every day, and 12 % at least once a week—based on a sample of 2,020 interviewees (cf. Statista, 2013b).
- **Social networks and blogs are the top category** by share of total online time. In 2011, US Americans spent 22.5 % of their Internet time on such websites (cf. Nielsen, 2011, p. 2).
- The **most important**—and thus most visited by far—**social network** is *Facebook* followed by *LinkedIn, Pinterest, Twitter,* and *Instagram* (see Fig. 4.4).
- The **use of social networks** among US adults (age 18+) differs depending on age, sex, and ethnic origin. In general, female adults spend more time in social networks than their male fellow citizens. The age segment of the 18- to 24-year-

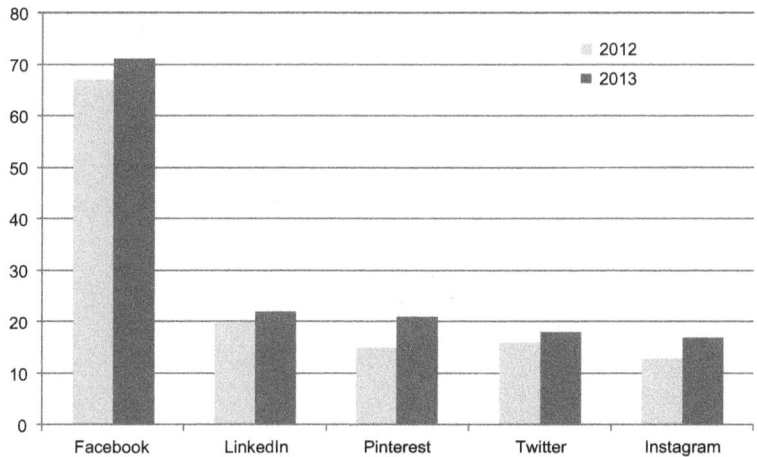

Fig. 4.4 Social media sites in the USA 2012–2013—% of online adults who use the following social media websites ($n = 1,445$ Internet users ages 18+) (*Source* Author illustration, data source: Duggan & Smith, 2013, p. 1)

olds dominates the use with regard to the time they spend per person on the PC in the social networks, but also the age segment of the 25- to 54-year-olds is strongly represented. When it comes to using social networks via mobile web and app, the 25- to 34-year-old US citizens dominate, followed by the 18- to 24-year-olds (cf. Nielsen, 2012, p. 6).

- This already shows that the importance of the individual social networks varies depending on the **mode of use**—as in at the PC, via mobile app, or web. At the same time, however, the list of the most frequently used networks hardly differs. The use of mobile applications is increasing more significantly than the use at the PC.
- *Pinterest* can display the **most significant growth rates** in use. In comparison to 2011, these are at 1.047 % in 2012 when used at the PC and 4.225 % when used via mobile web (cf. Nielsen, 2012, p. 6).
- *Foursquare* and *Reddit* are other social networks frequently used by US Americans, whereby they are represented most in mobile application (app or web; cf. Nielsen, 2012, p. 6).
- Regarding the **minutes spent per unique visitor** at *Facebook* in February 2013, the mobile usage (smart phones *iPhone* and *Android*) is with 785 min per month more important than that on the PC (320 min; cf. comScore, 2012; Kafka, 2013).
- The **most popular social activity on mobile devices** is among mobile social networking users to read posts from people known personally (80.1 %) followed by post-status updates (70 %) and read posts from organizations/brands/events (53.8 %) or those from public figures or celebrities (45.3 %; cf. ComScore, 2011, p. 23; cf. Fig. 4.5).

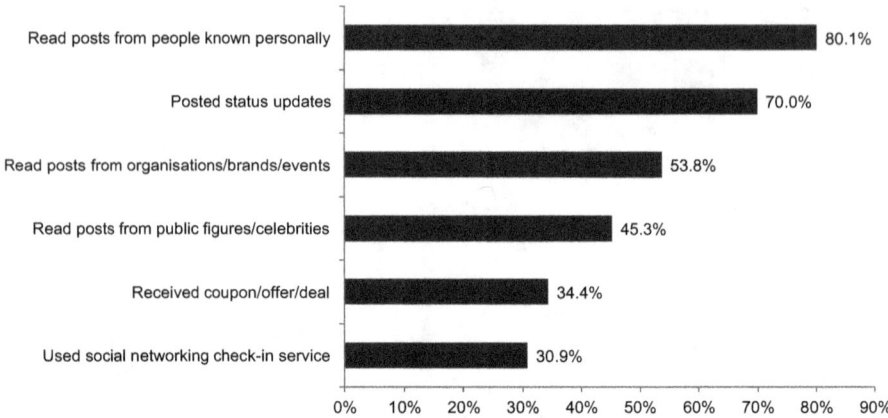

Fig. 4.5 Mobile social activities among mobile social networking users (basis: $n = 30{,}000$ US mobile phone owners, age: 13+; 3-month average: August–October 2011; representative sample) (*Source* Author illustration, data source: ComScore, 2011, p. 23)

A study by *Gigya* among more than 2,600 US adults (age 18+) in December 2012 shows that 53 % of interviewees have already adopted **social logins** for websites or mobile applications. Respondents also mentioned the reasons for bypassing social logins and the main benefits for users. Fifty-eight percent said that the main reason for using social login is the convenience, because users need not generate and look back on (new) usernames and passwords. Fifty-four percent% used such programs to avoid the completion of registration forms for every (new) social network, and 27 % mentioned to see a benefit in using social logins regarding **social syndication**. The main reason against the usage of social logins is for 49 % of interviewees because of the personal data transfer. Forty-one are concerned about the tracking and publishing of online activities without permission and 40 % of respondents bypassed such programs because of being not sure how personal data would be used (cf. Gigya, 2012, pp. 2-4).

One thing became evident based on the results presented: the **relevance of social media** and even the **social networks** in particular for companies can no longer be seriously questioned.

The relevance of the likes of *Twitter* today can be shown by the fact that *Barack Obama* made his first official statement on *Twitter* only a few minutes of winning the election in 2012: "Four more years." How the weighting of the media has been shifted overall is expressed well by a journalist of the *New York Times*: "During the 2008 election we were twittering what TV was doing. Four years later, television is talking about what we are doing on Twitter" (Rungg, 2012, p. 2).

Yet how many **companies** already have an **own social media engagement concept**?

According to a study that surveyed more than 3,800 marketers worldwide in 2012, the global share of companies using social media marketing is 94 %. Participants of the study came from different industries, countries—whereby the

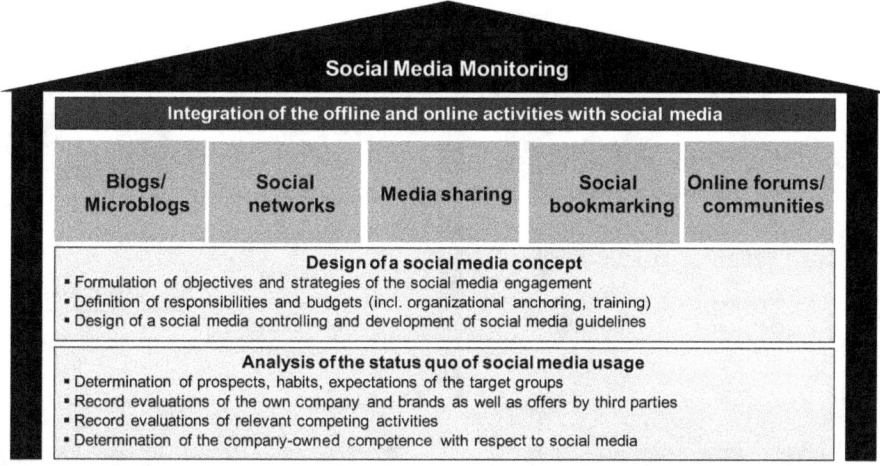

Fig. 4.6 The social media house—process of building up social media marketing (*Source* Author illustration)

USA is presented by 57 %—and sizes of companies (cf. Stelzner, 2012, p. 10, 41f.). An overview of the actual social media activity of the *Fortune 500* (also F500), a list of the 500 most successful—the largest businesses in size and wealth—companies of the USA annually compiled by the *Fortune Magazine*, will be given later in this chapter.

To make sure **engagement in the social media** doesn't become a flash in the pan, every company should work out a strategy for using social media before starting to. This also includes the provision of the necessary financial and human resources. The fundamental **process for a company entering social media** in general is shown by the **social media house** in Fig. 4.6.

A requirement for all measures is first of all an **analysis of the status quo of using social media** by the relevant stakeholders as well as the corresponding competitors. Here we need to capture which interests, habits, and expectations the own target groups have with regard to the corporate engagement in social media. In addition, the tonality and content of the evaluation of the own company as well as own brands and products made by third parties in social media need to be reviewed. The values determined here quasi-represent the **reference measurement of social engagement in social media**. For us, it represents an indispensible requirement in order to be able to ascertain possible changes in perception as well as in the evaluation in particular of one's own actions later on. Gathering and evaluating the activities of corresponding competitors in social media round off the status quo analysis.

When evaluating corporate social media competences, one problem area that frequently surfaces even with group dynamic processes must be considered. This is the **falling apart of the self-image and image of others**. The relevance of this contrasting can be demonstrated on the basis of the **Johari window** (named after

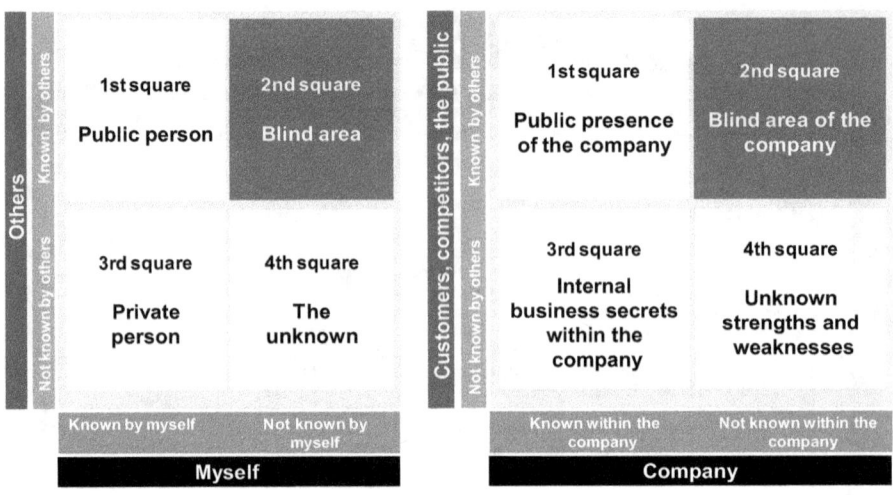

Fig. 4.7 Johari window for modeling interpersonal relations—oneself and the company (*Source* Author illustration)

the authors Joseph Luft and Harry Ingham; Rechtien, 1999, p. 95f.). With regard to self-perception and perception of others on a private basis, four squares must be differentiated (cf. left-hand diagram in Fig. 4.7). In the **1st square**, we speak of the public person, because it concerns behavior patterns and motives that are known to myself and my environment or which they can perceive. The **blind area** in the **2nd square** covers the behavior patterns others are able to perceive of me but which I myself do not know (e.g., long-established habits, speech quirks). The **private person** of the **3rd square** contains aspects that I know but keep hidden from others or don't want to be known. The **4th square** is reserved for the **unknown**, what I am not aware of and others do not know of. This is frequently termed the unconscious.

When translating this to everyday company life, the **1st square** reveals the planned and thus consciously scheduled **public presence of the company** to the outside and inside (cf. right-hand diagram in Fig. 4.7). The **3rd square** covers **business secrets within the company**, which are known internally and are applied for controlling the company, for example. These can and should be hidden from the outside. The unknown factors of the **4th square** include **unexploited strengths** such as certain employee talents which—especially in the context of social media—blossom in hiding. But they also included **unknown weaknesses** such as presenting the company on social platforms which previously have not stood out much in the company or on the market.

In the scope of the status quo analysis, it is particularly important to address the **2nd square** and thus the **company's blind area**. What do others know about us that is unknown to us? What do others see that we don't see? This can be an extremely bad image among a specific customer group that the company is not aware of. It can also be an "awful" quality of customer contact by phone or a very under-average quality of response on *Facebook*, about which everybody is speaking—possible even in public—only the company concerned isn't. The analysis described here should contribute to there being no "terra incognita" (as in of an unknown country or unexplored area of knowledge) in the 2nd square, but that we have ideally turned it towards the 1st square in a positive and value-increasing way or that weaknesses have been removed on the way there. This analysis can be fed with information by the instruments of **mystery market research**. These include mystery calling, mystery surfing, mystery shopping, down to mystery sleeping and mystery dining, in order to check the quality of service in hotels and restaurants. The stars awarded by the *Michelin Guide* are nothing other than results of such a mystery dining!

> **Think Box**
> - How can we determine the "blind spots" my company has?
> - Who can help us with the "unfiltered" look at our social media performance?
> - Who could be entrusted with such a task?

We are not able to tackle the **development of a social media concept** until based on the overall knowledge acquired here (cf. Fig. 4.6). For a start, this is where—in the old-fashioned manner—the **goals of the social media engagement** are drafted. In doing so, it is important to not run after external performance indicators—or even those defined by the own CEO—that are presumed such as the number of *Facebook* fans or *Twitter* followers. More important is the question: What is really to be achieved by engaging in social media? The core of the question is that companies first need to decide whether at all and, if yes, how they will engage in social media. In many cases, their potential customers and customers are already there and talking about the company, the brands, and/or concrete products. This needs to be determined in the course of the **status quo analysis**. Companies have three fundamental **options for action for social media marketing**. These can be described by the terms **listening** and/or **joining in conversation** (by reacting and acting):

- **Listening: Web monitoring**

    Listening by means of efficient **web monitoring** represents the **lowest level of social media engagement** that all companies should implement regarding social media, irrespective of their other Internet activities. It needs to continuously find out what is being said about the own performances in social media. For, even if there is no corporate involvement in social media, usually something is already being said, written or visually made public about the company and/or its products and services. The following applies to social media to an extent previously unknown:

▶ Social media never sleep!

For this reason, we must not restrict monitoring to classic working hours (incl. the weekend off); otherwise we possibly run the risk of dramatic developments beginning on Friday evening and in the truest sense of the word "slept through" by the companies concerned. Messages by individuals—whether true or not—could end up distributed over night or over the weekend at a viral speed without having been commented on and reach thousands, hundreds of thousands, or even millions of people (cf. the example on this of *Vodafone* in Chap. 9).

Web monitoring also provides us with the informational basis to develop the two following ways of using social media: "reacting" and "acting". For companies cannot simply barge into a "dialog" with their target group, but first have to determine where the target group engages, what moves it, and whether the own performances are being assessed in a positive or negative manner. To do this, the tools of web monitoring already spoken about are to be applied (cf. Chap. 3).

▶ **Remember Box** Long-term successful corporate acting in social media cannot be expected without actively listening—across all online and offline channels.

- **Reacting: Integration**

  Companies can step out of the passiveness of web monitoring and become **actively integrated in the communicative processes of social media**. This can become necessary if ongoing discussions there are not acceptable for a company. They may be false accusations, one-sided representations, or other disparagements against which the company wants to defend itself. For the following applies: "Web 2.0 harnesses the stupidity of crowds as well as its wisdom" (Grossman, 2006). In order to allow for this, the company can, on the one hand, take a stand on certain issues in its own name in blogs (like *Tumblr*) or via *Twitter* and try to influence the direction the communication goes in. On the other hand, companies can use existing platforms in order to post their offers and address their target groups in this way. This can be, for example, the setup of a fan page on *Facebook* or the provision of video messages on *YouTube* or *Vimeo*.

- **Acting: Creation**

  The most extensive way of engaging is the setup of own platforms in social media by developing own forums, communities, or blogs, in order to actively participate in the forming of opinions. This includes, among others, the setup of a corporate blog or the development of an own video channel on *YouTube*.

When it comes to the **utilization of social media** by means of social media marketing in the ways mentioned above, it is important to remember that the way, timing, and frequency of **using social media** is determined by the stakeholders themselves to a great extent. In the case of classic mass media, the use or to be more precise the possibilities of use are determined to a great extent by the communicating companies, media agencies, as well as publishers or the broadcasting companies. This is the case, for example, by the moment a TV spot is broadcasted; it is different, in contrast, with an advertisement. When it comes to social media, it may be the case that a company has decided in favor of *Twitter* but the Internet users suddenly start to communicate about the contents in blogs or on *Facebook*. This is where the companies must follow the users' channels to be listened to and express their appreciation of the stakeholders.

> Remember Box **The rules within social media are defined, monitored, and, where necessary, improved by the users.** Yet here, companies can make their own contributions and set the pace. A lever for creating the rules of the game to their advantage is something they don't have.

Yet what exactly are the **objectives of a social media marketing strategy**—geared towards the possibilities described above of listening, reacting, and acting to be aspired (cf. Fig. 4.6)? In many companies there is still a high level of uncertainty. First of all, it is essential to derive the social media objectives from the corporate objectives in a consistent way and not to try desperately to invent "totally new objectives" for social media! The "Marketing Strategy Report: Social Media" by *Ascend2*, conducted in February 2013, shows which **social media objectives** companies are aspiring today **worldwide** (see Fig. 4.8).

- **Improve customer engagement**
  The most important objective that 55 % of the respondents aspire with social media marketing is achieving improved customer engagement. Honest and constructive exchange between a company and its customers and among themselves ideally has a positive effect on customer engagement and thus on the overall awareness and reputation of a brand or company. By means of positive buzz marketing by dedicated customers, the ZMOT can arouse the interest of other users, their attention can be aroused, and the popularity of the offer increased. Social media is particularly suitable for viral marketing as in **online buzz marketing**. The communication from user to user can facilitate the viral distribution of contents in a particularly credible way via social media. Recommendations and links passed on by friends and acquaintances within the networks are seen—as already discussed—as particularly reliable. Such a viral distribution, however, not only takes place among friends but also among unknown third parties without losing much credibility. For this reason, we should all review how we can use our engagement in social media to increase the popularity of brands and companies.

  Besides the effect of positive buzz marketing hoped for, by improving the customers' engagement, a **relationship between the customer and the company** can be established or an existing one strengthened. The basis of this goal is

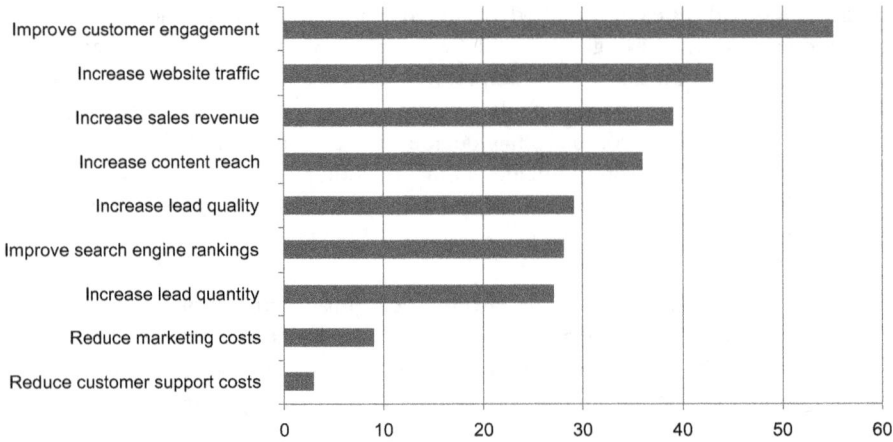

Fig. 4.8 Business objectives of social media marketing worldwide—in % ($n = 687$; question: "What are the most important objectives of your social media marketing strategy?"; multiple answers possible) (*Source* Author illustration, data source: Ascend2, 2013, p. 6)

formed by the knowledge that relationships to brands and companies nowadays—besides the promotional impulses, the experiences at the online or offline POS as well as the use itself—are increasingly affected by the **brand experience in social media**. It is surprising that this goal is not explicitly aspired by the companies questioned, for especially the establishment of relationships by **involving** the companies **in the information flows** and the **possibilities of a direct exchange of information in dialog** between companies and users represent the strengths of social media.

One way of improving the customers' engagement can arise in the cooperation with them to extend the product and services portfolio (**crowdsourcing**). The **use of wisdom of the mass** for innovation management within companies as an explicitly defined goal should also be reconsidered. In doing so, the involvement of collective intelligence of potential customers and customers can enhance various goals with the help of social media in the innovation process. On the one hand, customer-relevant innovations can be developed by means of creativity competitions. The creative contributions, however, do not need to be restricted to the product and services portfolio alone but can also cover the entire value chain of the company. In this way, suggestions for product names and advertising contents but also indications of attractive resources can be gained. On the other hand, such an opportunity to participate—even if only used by a small number of the potential customers and customers—has a positive **halo effect on the company image**. In addition, it can **increase brand and company awareness** and even **reinforce the acquisition of new customers**.

- **Increase website traffic**

    The second most important objective of social media marketing was stated by 43 % of the respondents as being an increase in website traffic. Subsequently, the participation in social platforms serves as a Trojan horse in order to draw the attention of relevant users to the company's website, brand, or campaign and to induce them to visit the corresponding website.

- **Increase sales revenue**

    Thirty-nine percent of the marketers interviewed stated an increase in sales revenue as an objective for using social media. This reflects the consequent deduction of social media objectives from corporate objectives already mentioned and recommended—at least for profit-oriented organizations. Furthermore, this explicit goal setting can be seen in the cause of the increasing pressure on **profitability of social media marketing** that those responsible for marketing are increasingly confronted with. In order to be able to justify corresponding costs, income, which can be attributed to the involvement in social media, must be generated. With that, the acquisition of new customers can be pursued or the quantity of the sales of customers (more-sell and/or cross-sell) or the amount of expenditure for offers (up-sell) can be increased.

- **Increase Content Reach**

    Thirty-six percent of the respondents stated an increase in the reach of the content communicated as an objective of social media engagement. Global networking via the Internet and in particular social platforms provide media with a previously unconceivable range of influence—*Facebook* on its own has more than 1.3 billion members worldwide! Furthermore, as already mentioned under the objective "improve customer engagement," via social media, content can be distributed more easily, faster and to a larger number of users by users themselves—considering the "retweeting" of posts on *Twitter* and the "share" or "like" function on *Facebook* and others.

    Combined with the aforementioned activities, the objective of supporting the **establishment of relationships to multipliers** should be explicitly followed. In doing so, it is essential that we may no longer place the **classic offline opinion leaders** (such as editors and journalists of established media) on the center stage of such activities. **Online opinion leaders**, who, as bloggers or twitterers with a large reader and/or follower community, can greatly influence the opinion on the web are becoming very important. These should be addressed in addition as important (digital) opinion leaders.

- **Acquisition of new customers and potential customers**

    After all, some of the companies interviewed are pursuing an increase in "leads"—i.e., new customers or potential customers via social media—by setting explicit goals. It is surprising that in doing so, an increase in the quality of new customers (for 29 % of the companies interviewed) is focused on more than an increase in the quantity (27 % of the respondents). For these companies, the use of social media is to establish direct contacts to potential customers. However, we should beware of redefining "social media" as "**commercial media.**" Using social media for direct sales will only meet with the users' great acceptance in rare cases. Social media rather offer communication "via connections," which should of course enhance sales, yet frequently rather indirectly.

- **Improving the company's search engines rankings**

    Twenty-eight percent of the companies interviewed are pursuing an **improvement in the rankings of search engines** by means of engaging in social media. By providing contents on various social media platforms, which are linked up to their own corporate website, the **search weight of the corresponding online contents** is increased significantly. We should exploit this possibility intensively in order to improve the findability on the web. Latest research shows that social mentions have a strong impact on the position in the organic listing (cf. Searchmetrics, 2013).

- **Reduce marketing cost**

    Nine percent of the companies interviewed stated that another important objective is the reduction in the costs of marketing activities. Yet we have to state that a company's engagement in social media means first of all additional costs, too.

- **Reduce customer support costs**

    Three percent of the respondents name the reduction of customer support costs as a social media objective. This constitutes the exciting possibility of offering and performing **(self) service features** via social media. It is thus surprising that there weren't more companies that stated this as an objective. And yet the function of social media can go far beyond being merely a complaint channel and also cover interesting aspects of pre-sales, sales, and post-sales phase in a customer-oriented way. The use of *Twitter* in particular is being increasingly recognized by companies as personal and direct customer service. A sphere of activity increasingly developed by companies in particular is the area **social service**. The public-oriented provision and rendering of services can contribute towards establishing a customer focus in the long run. As had already become obvious, customers practically force companies to demonstrate their service performance in social media (cf. Chap. 6).

The authors would also like to refer to further objectives at this point that companies can pursue with social media marketing:

- **Controlling the brand/company image**

    Exercising influence on the brand and company image can also be a social media objective. In doing so, we must remember that real "controlling" of the brand and company image is not achievable by engaging in social media. We can try to play out the contents determined within the brand and company identity via social media, too. Yet, there is no real possibility of controlling to achieve the image positions relevant for our brands and companies.

Participation in social media merely provides the opportunity to communicate contents that can contribute towards the establishment of a positive image for individual brands or for the entire company. If topics, companies, or products trigger emotions on social media platforms and occupy people intensively, such developments can quickly start rolling—both positively and negatively. By listening and in particular when companies actively participate in social media, problems or crises affecting the company can be learned from and, in a positive case, **PR catastrophes** can be averted. Companies can try in this way—under their own name—to actively fight and rectify negative opinions. At the same time, they can reinforce positive views and "reward" those who support the brand. If the CEOs, CMOs, or managing directors also avail themselves of social media to communicate with the various stakeholders via *Facebook, Twitter,* or in corporate blogs, this can ideally have a positive effect on the company's image.

- **Market research and market observation**
  The social media objective of market research and marketing observation should also be defined. After all, expressions of opinions in social media provide an unhindered and unvarnished view of the evaluation of the company's performances from the stakeholders' point of view. We should make 100 % use of this **chance for acquiring information**—knowing that only a (inconsistent) random sample of our customers get involved!

- **Acquiring new employees**
  Companies should also consider using social media for the acquisition of new employees and accordingly express this as an objective. When doing so, we must be aware of the fact that the previously addressed ZMOT also exists with regard to employment relationships. After all, more and more information from an internal perspective of the company can be found on the Net—and hopefully not only from frustrated former employees! We must not neglect this aspect. And also for this reason, listening as in market research and market observation is something we as companies cannot do without.

Think Box
- What objectives form the basis of our social media marketing?
- Have these objectives been expressed in writing?
- How systematically are we involving social media to increase the publicity of my company and our brands?
- Is the social media objective of "acquiring new customers" a dominating factor for us? How do we monitor the corresponding achievement of the objective?

(continued)

- Do we use social media engagement to increase the quality of hits on search engines?
- Are we keeping our eye on the online opinion leaders in terms of communication and are we providing them with exciting information?
- To what extent do we use social media for market research?
- Are we exhausting the potential for acquiring new employees for social media?
- How widely or narrowly is our search pattern defined, in order to use "wisdom of the mass" for my company?
- Is the achievement of our social media objectives checked based on informative KPIs?
- Do these KPIs bear relation to the monetary objectives of my company (turnover, profit margin, EBIT)?
- Are remuneration systems for executives and employees linked up to such objectives?

It is fascinating that companies—based on the study by Ascend2 (2013)—have not yet recognized that social media can also be involved in informing their **own employees** about visions, values, objectives, strategies, as well as about ongoing campaigns and events. The use of blogs, wikis, but also social networks themselves can make a significant contribution towards the provision of information from "top to bottom" but also from "bottom to top" as well as between various areas and departments—even beyond country borders. Social media thus become an important building block of **intra-company knowledge management** (keyword: "social intranet").

In order to exploit the **potential of social media**, the perspective should be additionally expanded at this point. Even if the study by Ascend2 (2013) shows that the majority of companies primarily try to achieve communication or promotional objectives via social media, their potential is much greater. Figure 4.9 provides a view of how **social media can be used in the value chain of a company**. An important area of performance of social media already discussed in Chap. 2 is the **involvement of users in product development**. A key aspect, which also became clear in the discussion about social media objectives in Fig. 4.9, is represented by the **communication on social media**, which is possible via and with companies. A further important sphere of activity is represented by **social commerce**, which will be discussed in Chap. 7. In addition our service can become "social," too, by offering service via social media platforms or by integrating own customers in service delivery (e.g., by peer-to-peer service platforms).

We are well advised to keep this entire value chain in mind if we want to recognize the potential of social media in their depth. Then **social media** becomes the **driver of the entire value chain**.

4 How the Social Revolution Is to Be Managed

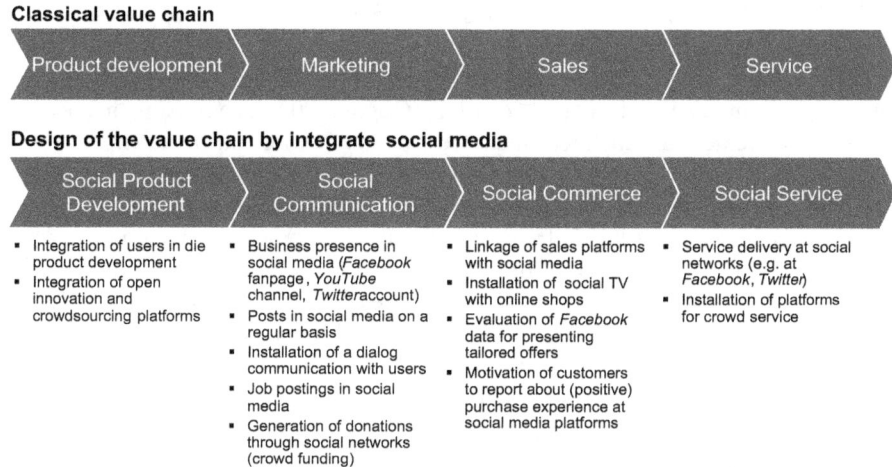

Fig. 4.9 Integration of social media in the value chain of the company (*Source* Author illustration)

> **Think Box**
> - Have you ever checked to what extent we can use social media to communicate with our own employees?
> - Which other areas of our value added chain can we "pollinate" via social media?
> - Which possibilities do social media offer us to improve the quality of our service with an effect on the public?
> - Who within my company can systematically answer these questions?

When we decide to involve social media in our dialog with the relevant opinion leaders of our company, with our potential customers and customers, as well as with other stakeholders, we should consider the following **basic principles of communicating in social media**:

- **Honesty/authenticity**
- **Openness/transparency**
- **Communication at eye level**
- **Relevance**, e.g., by means of context/location orientation
- **Continuity/sustainability**

Honesty and **authenticity** (to be understood as "genuineness") in communication represents a key basic principle in social media. If we tried to influence the formation of public opinion for our own sake, by making up our own evaluations and comments about ourselves, we would be taking a great risk. If such behavior were uncovered—and the probability is extremely high given the many online specialists with extremely high time budgets and an affinity for investigating—our image can be (sustainably) damaged. What has applied for a longer time is: "Today, there's nowhere to run and nowhere to hide. The moment you hide something, you will end up being exposed and picked apart" (Gogoi, 2006). A company that uses or has used, for example, unethical measures or that cannot fulfill the expectations built up by its customers should be discouraged from entering social media. If a company has "skeletons in the closet," which could be disclosed by active Internet users, information about this is frequently virally distributed. Such a company has already established the potential for a shitstorm.

The engagement in social media presupposes as a further basic principle the ability to accept criticism from all kinds of stakeholders as well as to respond to it openly and authentically. Only by means of **openness** and **transparency** in regular communication with all kinds of stakeholders are we able to achieve the high amount of credibility necessary. The opposite will be achieved if we don't become visible as participants in social media until erroneous information, which we wish to rectify, is already being circulated. The messages then communicated frequently do not pass the "smell test," because we had not yet managed to become integrated and established in the social media sphere. Longer-term engagement in social media in contrast holds the necessary communication channels ready—even in cases of crisis.

> ▶ **Food for Thought** "Transparency may be the most disruptive and far-reaching innovation to come out of social media."
> *Paul Gillin*, social media expert

When it comes to dialogs and discussions—not only, but in social media in particular—we should seek communication at eye level as a further basic principle (cf. Fig. 4.10). The patronizing, (presumably) better informed company or its representative, in contrast, will hardly meet with acceptance. This applies even more for every type of monological communication. With every inquiry, every contribution to a dialog in a community, forum, or blog, before reacting to it, we should first presume that a well-networked communicator may be behind it. We should rather face them, and not only them but all the others in dialog with appreciation and respect. If the corporate engagement in social media is not assuring, this can lead to a so-called backlash and thus to a deterioration in the acceptance of brands, offers, and/or the company as such. If companies do not adapt themselves to the changing conditions and rules of social media, there is an increased risk of failing. Maybe the time will even come when the customers—as already mentioned—become the **masters of communication** and dominate communication.

Within social media—like otherwise in life—it is all about winning the attention of the target persons. The key term for this as another basic principle of communication in social media is **relevance**—irrespective of whether we try to present our own offers, whether we invite users to participate or are invited by them to participate. A requirement for engaged participation is that the contents and offers

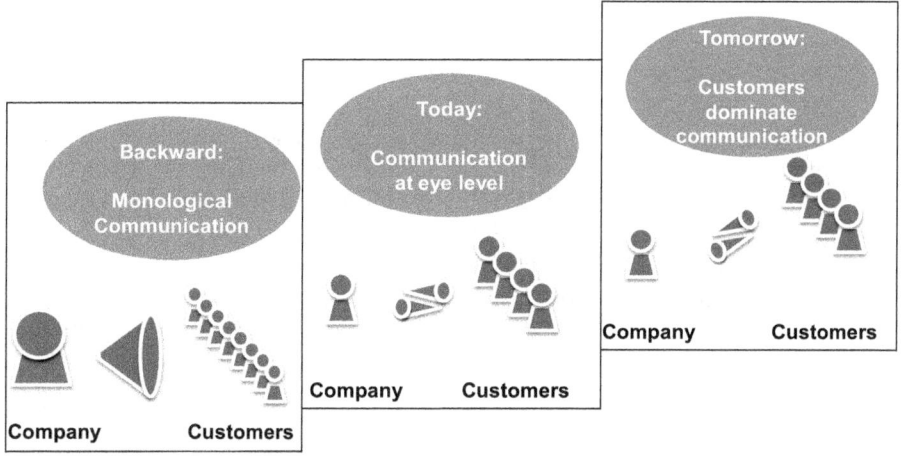

Fig. 4.10 From a monological communication to a communication at eye level and finally to a communication dominated by customers (*Source* Author illustration)

presented are of relevance to the target groups. The overall objective of companies should thus be to establish a long-term relationship based on mutual appreciation, loyalty, and trust by means of various types of interaction with the users. And the requirements for this to succeed are contents that are interesting. It is good to anchor a downright **passion for relevance** in the entire organization to the greatest extent possible in our company (cf. Chap. 9 concerning the digital transformation).

A further basic principle for communication in social media is a minimum of **continuity** or **sustainability**. If fans and followers are gained and users engage in our blogs or forums and provide user-generated content as earned media, then our social media engagement must not be a mere flash in the pan. This is the reason why we should only start such campaigns (invitation to join in campaigns, the announcement of events) within social media where we can also secure ongoing engagement. This is where the well-known motto thus applies: Making enthusiastic on its own is not enough! One must also be able to deliver! For this reason, the entire organization is to be geared towards the integration of social media before larger steps in using it are taken.

The overall **credibility** offered by companies, brands, and offers can only be achieved if the communication in social media is consequently geared towards the named basic principles of honesty/authenticity, openness/transparency, relevance, and continuity/sustainability and if, at the same time, communication is at eye level. It is not until then that the ever increasingly important **trust** of our business partners in us will set in—a currency that will dramatically increase in importance (cf. Chap. 6).

> **Think Box**
> - How honest and authentic is the communication we carry out in social media?
> - How successful are we in achieving openness and transparency in the communication about the company as well as our brands and offers?
> - Does an outsider experience our communication at eye level? Or do we still patronize and instruct our stakeholders to a great extent?
> - How successful are we in providing really relevant contents? How well established is the "passion for relevance" in my company?
> - Is our engagement in social media geared to be long term or is there a danger of a flash in the pan?
> - Who monitors that the overall basic principles of communication in social media are observed?

In order to also actually reach the defined objectives with our engagement in social media, we need to draw up a **social media strategy** before entering social media (cf. Fig. 4.6). This also includes the provision of necessary financial and human resources as well as the type of organizational anchoring incl. the development of a social media controlling system as well as social media guidelines. Another particularly important factor is the CEO support, because the extensive exploitation of social media potentials goes hand in hand with a change management process (cf. Chap. 9 concerning the digital transformation).

Similar to customer retention systems, the various **offers in social media** first of all attract the fans or those persons who have now already established the closest connections to the company and its brands. If an offer is then removed from *Facebook* or *Twitter* or *Pinterest* after a short period, the contacts to the most important partners of the company may possibly be capped. For this reason, when entering social media, an **exit strategy** should always be contemplated. What this means precisely is that the first engagement in *Facebook*, for example, is initially a "six-month test phase." If the corresponding activities are then ceased due to a lack of achieving the objectives or resources, nobody should be surprised. If, in contrast, the activities are continued, there should be no loud protest.

> ▶ **Remember Box** Every company is well-advised to anticipate a possible exit strategy before entering social media marketing.

However, it is sobering to take a look at the **requirements for implementing social media marketing** which have been created in companies. These can be suggested with regard to the challenging **obstacles regarding social media activities** among businesses worldwide revealed by the Ascend2 study (cf. Fig. 4.11; cf. Ascend2, 2013, p. 7). In doing so, it becomes evident that 42 % of the companies **lack the necessary employees**. Thirty-eight percent are not able to **measure the return on investment of their social media** marketing strategy. Twenty-eight

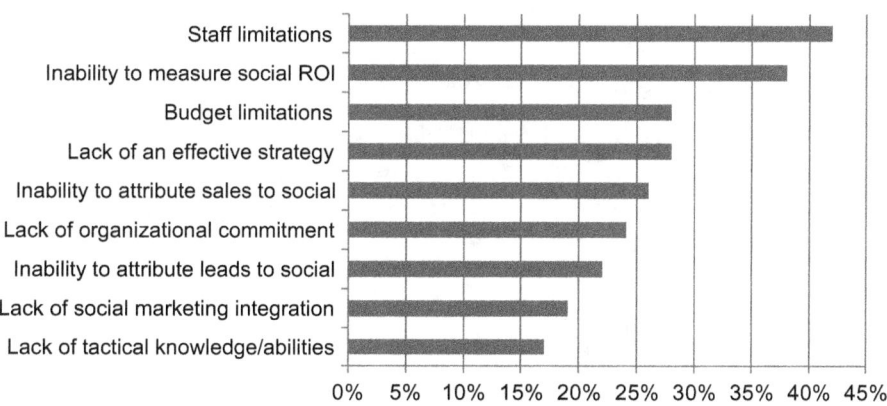

Fig. 4.11 Most challenging obstacles to reach social media marketing objectives by businesses worldwide—in % ($n = 687$; question: "What are the most challenging obstacles to achieving your social media marketing objectives?"; multiple answers possible) (*Source* Author illustration, data source: Ascend2, 2013, p. 7)

percent each state having **too low a budget** or that **no effective strategy** was laid down. In 26 % of the companies, it is **not possible to ascribe sales and** in 22 % "**leads**" to the activities in social media. Twenty-four percent see the **lack of organizational commitments** as one of the greatest challenges for implementing social media activities. A **lack of integration of social media activities** is stated by 19 %, and the **nonavailability of knowledge or skills for social media** activities are stated by 17 % as a challenging obstacle to reach social marketing objectives.

All in all it becomes obvious how few companies are prepared for the **challenge of social media** and why success does not want to materialize in many areas! This explicitly illustrates the gap between individually defined social media marketing objectives and the metrics for tracking success (cf. Fig. 4.12; Ascend2, 2013, p. 11).

Based on our findings regarding the objectives and strategies of social media activities, the suitable instruments and platforms of social media must now be selected (cf. Fig. 4.6). The most important factor in doing so is the question as to whether our company offers enough substance to deliver attractive and thus relevant contents for the various stakeholders. No social media engagement will be successful without convincing substance and thus without a convincing **content strategy**. Sometimes **content marketing** is already spoken of—as if there had ever been a successful marketing strategy "without content." At any rate it is essential that "Content is king!" and not only the reach counts. The latter is merely the necessary requirement for successful communication but, on its own, is not adequate for the social media user "to stay on the ball" and for the defined social media objectives to also actually be achieved.

In order to present the **social media activities for US-American companies**, the results of a study by the University of Massachusetts Dartmouth should be presented at this point. It studied the engagement of the largest companies of the USA (Fortune 500) with regard to their **engagement in social media**. In 2012, 73 %

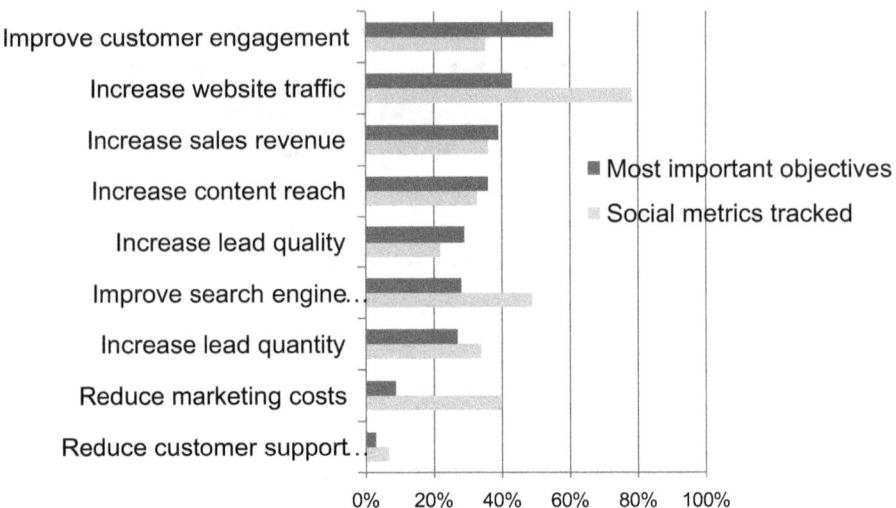

Fig. 4.12 Gap between defined social media marketing strategy objectives and the tracked social media metrics by businesses worldwide—in % ($n = 687$; results compared answers of the question "What are the most important objectives of your social media marketing strategy?" with the answers to the question "Which of the following social marketing metrics does your company track?"; multiple answers possible in each case) (*Source* Author illustration, data source: Ascend2, 2013, p. 11)

of the F500 had a corporate *Twitter* account—whereby in 2011 it was only 62 %—and 66 % had a corporate *Facebook* page, representing an increase of 8 % compared to the year before. In 2011, 31 % of the companies had neither a corporate *Twitter* nor a *Facebook* account, whereby in 2012, it was only 23 %. In 2012, the engagement in the video sharing platform *YouTube* was studied for the first time. It as ascertained that 62 % of the largest US businesses have a corporate channel. In addition, 28 % of the F500 have a corporate blog meaning an increase of 5 % in comparison to the year before (cf. Barnes, Lescault, & Andonian, 2012; Fig. 4.13).

With regard to these social media activities, it must be mentioned that the **industry** in which the respective company operates and the **rank within the F500** list are of significance. Thus, the ten largest US companies, for example, have a corporate *Twitter* account and post regularly, whereby eight of these also have a corporate *Facebook* page. Of the 200 largest US companies, 44 % have a corporate *Twitter* account, 42 % a corporate *Facebook* page, and 54 % even have a corporate blog. Companies from all 71 industries have a corporate *Twitter* account but *Facebook* activities can only be noted by companies of 69 industries and companies from only 54 industries even have a corporate blog.

Besides the presence of the corporations, a small number of "**specialty accounts**" or **blogs** can be noted among the F500. These focus on special topics such as information about employment/career at the companies. Eighteen *Facebook*, 17 *Twitter*, and four *YouTube* accounts could be counted among the F500 companies as well as seven special blogs. Some companies are still using new

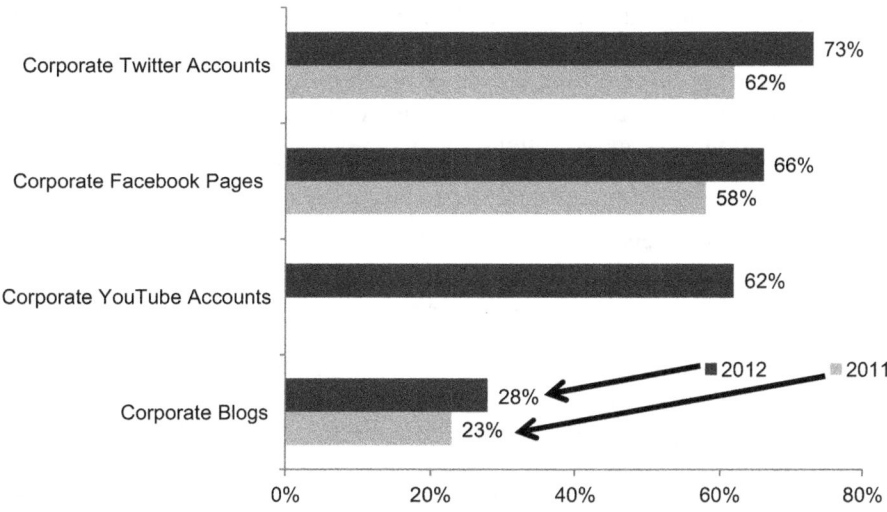

Fig. 4.13 Major engagement in social media of the Fortune 500 companies (*Source* Author illustration, data source: Barnes et al., 2012)

social media tools to establish contact with their stakeholders. Yet up to now, merely about 2 % of the F500 are active via an account on the content sharing platform *Pinterest* (cf. Barnes et al., 2012).

The results of a study by *Awareness* also reveal a similar picture. Besides the social media activities of large companies, they also present the smaller ones. Thus, 47 % of the companies interviewed generated less than a million US $ revenue. A total of 89 % of the companies interviewed stated being active on *Facebook*. *Twitter* is named by 84 % and *LinkedIn* by 77 % in second place. The video sharing platform *YouTube* was used by 71 % and blogs by 64 % of the companies interviewed in 2012 (cf. Awareness, 2012, p. 16, 39). Furthermore, the content sharing platform *Pinterest* is increasingly used and other used social platforms are forums, *Flickr, Slideshare, Foursquare,* and *Tumblr* (cf. Awareness, 2012, p. 36). In this study, it becomes more obvious than in the study of the social media activities of the F500 that companies increasingly count on multiple accounts than on just one platform, in order to more aptly address certain segments. Fifty-three percent of the companies interviewed stated having more than two *Facebook* accounts and 13 % had even more than five. Forty-five percent of the respondents have more than two accounts on *Twitter*, whereby a total of 11 % had more than five (cf. Awareness, 2012, p. 17).

Furthermore, this study reveals that other social technologies are being increasingly used. Thus 65 % of the companies interviewed state using **community platforms** in addition. Thirty-three percent of the respondents used **collaboration platforms,** 23 % **social commerce platforms,** and 15 % **innovation platforms** within their social media engagement concept (cf. Awareness, 2012, p. 31).

When **developing** and in particular when **implementing a social media concept** (incl. the organizational anchoring as well as training the employees), attention must be paid to not only reaching a target group-oriented **networking of individual social media** but that **networking with other communicative measures** of the company is achieved, too. In doing so, it is essential to organize the entire communication mix and thus in a cross-media way into the following three areas:

- **Establishing awareness**
- **Securing engagement**
- **Achieving a conversion** (as in a purchase)

Ideally, this form of networking takes place at as early a stage as when defining the social media objectives as well as when laying down the social media strategies. This is the only way to create a coherent overall image of the company which is indispensible for achieving the marketing objectives. The entire social media engagement must be incorporated in **social media monitoring** to find out the desired and undesired results at an early stage and comprehensively (cf. Fig. 4.6). In order to be able to act fast, the significance of **real-time monitoring** is increasing to have more time to act. The basic concept was presented in Chap. 3.

Many companies still abstain from using social media because they are **scared of losing control** over their communication and performances. In all fairness, it must be admitted that the companies have long lost control due to the diverse possibilities of Web 2.0! Accordingly, corporate engagement on social media platforms is about partially compensating and/or moderating the loss of control, in order to not be completely excluded from the **game of social media**. This is indispensable for those companies in particular that have become very important for the public, their potential customers, customers, and other stakeholders. For, due to the **range of influence of social media**, negative statements or scandals can be distributed faster and damage the image in the long run—particularly if the companies are not present and do not react competently. The minimum engagement in social media thus represents the previously mentioned **monitoring** of messages exchanged there to see how companies, offers, and brands are discussed and presented. Social media in particular offer a never before imaginable possibility of having one's "hands on the pulse of the target group" and to learn in real time what they are interested in. However, if such monitoring is not done, no countermeasures can be initiated in a timely manner in the relevant media.

The fact that it is not easy for us to manage this task is something we must be aware of. For the **mass of users** changes its opinions, recommendations, and/or behavioral patterns fast, with no warning and not necessarily logically reasoned. The **virility of the expression of opinion** can however—nobody asks whether rightly or not—rapidly "infect" large user circles. The **instability of opinion** often makes it difficult for companies to act in social media. In addition, information distributed online can hardly be removed from the Internet. Therefore, crises are not only spread more rapidly but also more sustainably than it would be possible offline, if the company does not react at as early a stage as possible and adequately.

A prompt reaction by the company to negative comments is indispensible, in order to prevent the distribution and negative image of the brand.

> **Remember Box** A company has hardly any possibilities to control its image in social media in an extensive way. Theoretically speaking, a company can do without **online reputation management**—but not without an **online reputation**. Yet the question is whether the latter is significantly influenced by the company and features the desired content or whether the company is "driven" by users. The opportunity of social media is, at the same time, still disregarded due to a dominating anxiety in the companies: companies now have the chance to become actively involved in discussions on these platforms!

What **risks** for companies can be involved in an **unsystematic entry into a social media engagement** is shown by the many hits coming up on *Google* when entering the search terms **social media fails** (2014 approx. 190 million hits). In many publications, however, it is primarily emphasized which companies acted particularly successfully. This perspective, however, does not do justice to the complexity of social media. It is imperative that we also become occupied with fails in social media, as this is the only way we can overcome the so-called survivorship bias. What is behind that? Many analyses and "guidebooks" concentrate on the "winners" of the various competitions. These only show what superiority certain approaches have—based on examples of "winners." In how many cases a similar approach has led to failing remains unconsidered. This is why the results and thus the chances and risks of various concepts regarding these winners ("survivors") are distorted. We can counteract such a distortion by considering those in the analysis who did not make it and who thus did not count among those who succeeded. By doing so, the results are frequently shifted.

> **Remember Box** When analyzing social media activities, as well as with other analyses, it is important that we do not merely concentrate on the successful examples. And it is at least just as important to analyze the approaches of those who failed, in order to evaluate the chances of success of concepts, for we can learn a lot from losers! Anyone who wants to be permanently successful should take a look at the losers of a game, too.

In order to prevent such a survivor bias in social media, we have analyzed a large number of companies that were not granted success in social media. The most important **factors of failure when engaging in social media** that were discovered are identified in the following:

- **No self-critical analysis of the status quo**

  In some cases, companies were surprised about the **aggressiveness of the reactions** caused by the first engagement in blogs or in social networks. In these cases, the company or the responsible executives were not aware of the extent to which **dissatisfaction among their own clientele** prevailed. This has now for the first time found a visible valve for the company, but also for the public interest. Becoming aware of the previously mentioned "blind spots" in such a way can be particularly painful and is something we should definitely try to avoid.

  > ▶ Remember Box An **extensive status quo analysis** regarding the evaluation of one's own company by stakeholders is an indispensible **requirement of any social media engagement**.

- **No online response in the event of an attack online**

  Sometimes, companies that are attacked in social media try to justify and defend themselves by means of the classic (and trusted because learned) offline media. By doing so, the fact is neglected that the initiator of such attacks frequently cannot be reached by classic media, thus giving the attackers room for action where the company affected does not act. For this reason, we as a company must almost imperatively accept the **communication channels** defined by the critics, even if we are not that familiar with them. If a negative video is posted on *YouTube* and it communicated further via *Twitter* and *Facebook*, then the company's engagement in these three media is required to reach the sender and recipient of the negative mention. This also makes it obvious what the statement "dialog at eye level" means.

  > ▶ Remember Box The **dialog within social media** must always **be pursued where positive or especially negative mentions are circulating**. In this case, they must almost imperatively follow the user community—and not switch over to communication in a medium that is possible more up the company's "alley" (e.g., classic press releases or press conferences).

- **Using tedious legal defense instruments on online attackers**

  Using legal means against—from the company's point of view—unjustified attacks frequently leads to nothing. First and foremost, the people or institutions behind the "disagreeable or wrong" statements cannot be traced. Secondly, taking **legal action** is often drawn out so that the attack has possibly already run dry, and it has already achieved its long-term damaging effect before aspired judgments have been made. Thirdly, a courtroom dispute frequently leads to other media and users becoming aware of such a dispute. In many cases, a warning or the announcement of proceedings—with which we do ourselves and our company a disservice—suffices.

> Remember Box As a rule, users interpret **taking "legal action"** as a sign of weakness and doing so is punished in the media. A permanent solution that promotes the reputation of the company is rarely or never represented by this step.

- **Company-guided misusage of review platforms**
  Even if companies find it tempting to praise their own performance on various review platforms, this type of manipulation should be warned against. There are enough Internet users that want to make their mark by uncovering precisely such **manipulations** and make them visible to all in relevant blogs. In plain English, this means we should refrain from only expressing good feedback and having bad feedback classified as of little help. On the other hand, companies motivating satisfied customers to post their assessment on such platforms cannot be taken amiss.

  > Remember Box Companies should consistently **refrain** from putting themselves in a better light by means of **self-written posts and feedback** on review platforms. This is frequently recognized and punished very consequently.

- **Placing paid PR contributions in relevant communities, forums, and blogs**
  Internet users can identify manipulated or paid posts in communities, forums, and blogs or those added on purpose under a different identity much easier than is the case in classic media. An appropriate approach by the *Deutsche Bahn*, with which the privatization by means of positive articles in newspapers, readers' letters, and blog entries (e.g., on *Spiegel Online*) was to be supported, was uncovered and caused the redundancy of the employees responsible for it.

  > Remember Box **Refraining from falsified or paid PR contributions** in social media is indispensible for a convincing online reputation. Cheating in social media won't get you far.

- **Great time delays between "attack" and "counterattack"**
  The fact that the Internet is a fast medium for spreading messages by viral effects particularly rapidly should really be given extra mention. However, companies still react to attacks on the web far too frequently with **great time delays**. The longer unjustified or even justified criticism can be found on the Internet to which a company does not react, the greater and more sustainable are the damage to the image. In addition to that, the motto "felt" is: A non-defender is in the wrong!

  > Remember Box **Speed of the reaction** to user comments in social media is a given! It must thus be noted: One cannot not communicate! (*Paul Watzlawick*). This means that even a company remaining

silent is a statement which, in turn, can be extensively (negatively) commented.

- **No appreciative reaction to online comments**

    Even if companies have reason to criticize the contents and form of many online statements and contributions, they should take their critics seriously and react to their statements in an appreciative manner (even if it may sometimes be difficult). Reacting in an arrogant, ironic, or patronizing manner can cause a **communicative disaster**.

    Sometimes, a non-reaction to a "polemic attack" can be better. Before we remain silent, however, we should scrutinize the direction in which the communicative wave is moving and whether there are fears of expecting a **topical career** on the Internet. The cases in which there should be a reaction depend, of course, on the **significance of the communicator**. If this is an unknown person with a small network, it may be more appropriate to remain silent than if an active, generally known blogger or twitterer having an intensive network makes a critical statement. For this reason, the **degree of networking of the critics** should be found out.

    ▶ Remember Box **Appreciation** is the leitmotif for all communication in online media—and not only there!

- **Adopting content from other sources without verification**

    In order to create an up-to-date and dynamic website, in some cases, companies have adopted contents (such as *Twitter* tweets) containing the name of the company or own products and services (unfiltered) for promotional purposes on their own homepage. Of course, this poses great risks. If opponents or "jokesters" wish to evade this mechanic, then all they need to do is to post appropriate contents with the relevant names, in order to appear on the corporate website. There are no limits to the creativity as to how such could turn out to be!

    ▶ Remember Box One should **refrain** from an **(unverified) adoption of contents**—in particular with online sources. Despite other expectations, care before speed.

- **Using unsustainable and/or clearly refutable statements in corporate communication**

    It should be a matter of course that untrue statements should be refrained from in corporate communication. Whereas such statements in the past frequently remained uncommented because there were no powerful media to about them, the critics of today avail of a variety of media platforms. Some Internet users systematically look for counterexamples to official corporate statements and uncover, for example, which companies only operate greenwashing instead of actually confronting social responsibility. By **greenwashing**, attempts are made to establish the image of an ecology-conscious company in public, without undermining these by corresponding activities.

Such behavior becomes particularly dramatic when—as already has happened—corresponding *Facebook* groups for such campaigns are "entered" and the contradiction between saying and doing is pointed out. Sometimes, websites are also set up to conduct **negative campaigning**, which is colloquially also called "mudslinging." This generally means the attempt to gain superiority over competitors by communicating negative statements about them. In the context of social media, negative campaigning can be used by, for example, noncommercial organizations such as *Greenpeace* but also by individuals, in order to spread errors by companies against the own corporate values and behavior codes. By doing so, the affected companies should be motivated or forced to correct the course of their behavior patterns and other companies refrained from "wrong behavior."

> Remember Box **Honesty** is an important general principle for corporate communication—not only in social media. Dishonesty can be uncovered especially in social media and can be very easily denounced worldwide.

- **Insufficient integration of the various social media engagements**
  Internet users are frequently active in various social media at the same time—frequently expect this from their "corporate partners," too. For this reason, social media engagement should be systematically geared towards integration. For example, a *Facebook* page can be linked up to a *Twitter* account to communicate own tweets. If the corporate blog and/or website have buttons for the social bookmarking platforms, corresponding contents can be evaluated and—ideally—be recommended via these platforms. Videos on *YouTube* or *Vimeo* can be integrated in blogs like *Tumblr*, on *Facebook* pages, and in the own website.

  > Remember Box An engagement in social media stands or falls with the extent of the **integration in the overall corporate communication** or the entire company presence.

The **social revolution** is in full swing. It is up to us to check how we should position ourselves within these altered framework conditions, in order to safeguard our **corporate survival**.

> Food for Thought The **social landscape** is becoming more challenging. Are we ready and able to develop into a **social enterprise**?

Quick Wins

- _____
- _____
- _____
- _____
- _____
- _____
- _____
- _____
- _____
- _____
- _____
- _____
- _____
- _____
- _____
- _____
- _____
- _____
- _____
- _____
- _____
- _____
- _____
- _____
- _____
- _____
- _____
- _____
- _____
- _____

How Marketing Becomes the ROI Driver Within the Company 5

In Fig. 1.2 it already became clear that 56 % of the CMOs felt insufficiently prepared to take over the **responsibility for the ROI**. In order to overcome this deficit, important benchmarks will be conveyed in this chapter. A key term to this end for the entire company and for marketing in particular is the customer value. It is imperative to recognize the customer value in both of its perspectives of significance: On the one hand, it is the value a company provides its customer and motivates him to make a purchase (hereinafter "**value for the customer**"). On the other hand, it involves the value a customer generates for the company (hereinafter "**customer value**"). Both value contributions must be put into balance because otherwise, no company can survive in the long run (cf. Fig. 5.1). As a consequence, the important **do ut des principle** (i.e., "I give so that you might give") must be taken into consideration with regard to the balance between the "value for the customer" and "customer value." The equation thus reads as follows: the customers develop the greatest "customer value" for us when we generate the greatest "value for the customer."

One main reason why CMOs feel badly prepared for **taking over responsibility for the ROI** can be seen in the **lack of focus on the customer value**. In many cases, the task of marketing was seen to be creating added value for the customer as in the "value for the customer" and attending in full to his expectations—especially when dealing with social media. At the same time, the question was asked less clearly and less loudly—even by marketing and those responsible for it—about what (additional) customer value was to be achieved by the various activities. The question about additional budgets can be posed fast—the answer to customer values to be achieved in addition, in contrast, is much more difficult to provide.

How important is the **discussion about the customer value** or the **focus on the results of marketing** in general today—even in our company? All in all, it must first be ascertained that there is frequently still a lack of focus on consequent results in marketing. In many cases, such a focus is insufficiently pronounced. Yet, it is indispensable for a wide acceptance of a company's marketing if marketing and those in positions of responsibility wish to leave the corner of the "cash burners."

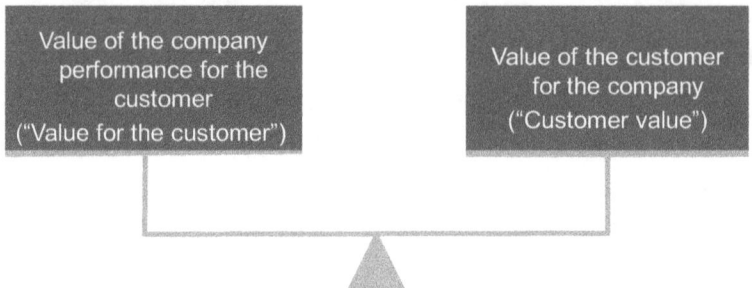

Fig. 5.1 Ensuring a balance between generated value contributions (Source Author illustration)

This means nothing other than that the CMOs should make an active and extensive effort to make their contribution towards achieving monetary company objectives in particular visible and thus also assessable. The magic formula for this is **Return on marketing investment** (ROMI) as a special form of the ROI. This is why it is essential as early as at the stage of the conceptualization of marketing measures to ensure that **control points for monitoring success** are included and meaningful **key performance indicators** (KPIs) are defined. An own study conducted in 2010 of companies already employing dialog marketing and thus being analysis-affine revealed, however, that merely 43 % conduct a **profitability analysis of marketing campaigns** relating to campaigns (cf. Fig. 5.2). Detailed evaluations on the level of product offers, products, or advertising means are carried out even more rarely. Yet the following still applies: questions about measurable results in marketing, online marketing, and particularly with regard to social media are relevant—but frequently unpopular!

A **value-oriented customer management system** represents a specific development of the focus on the results of marketing which relies on the customer value as a key performance indicator.

> ▶ Remember Box By means of a **value-oriented customer management,** marketing can demonstrate its monetary contribution towards achieving company objectives in the long run.

In the majority of companies, however, this **customer value** still does **not** represent a **key control tool**, as also shown in the study just presented. According to it, companies analyze measures such as the acquisition of potential and new customers primarily on a campaign level (43 %), whereas an evaluation of customer groups or on a single customer level represents a great exception with 23 and 18 %, respectively (cf. Fig. 5.2). Not much has changed since this study was conducted, as many interviews with marketing managers showed. This makes it clear why there are still such great insecurities regarding the necessary **acceptance of the responsibility for the ROI**. The **determination of the contributions of marketing for the company** frequently stops at **potential-oriented objectives**. Yet, it is all about

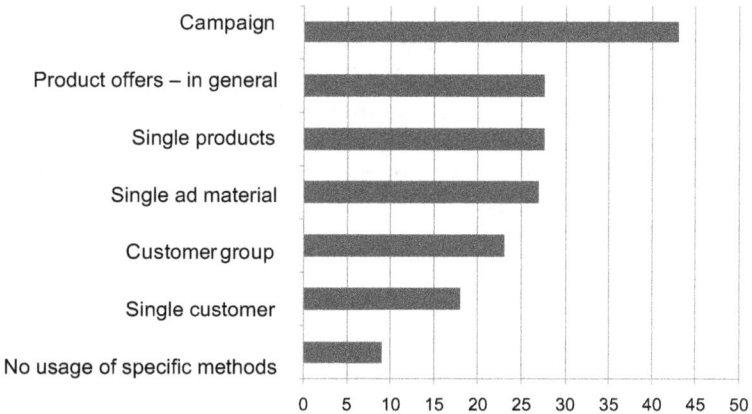

Fig. 5.2 Levels on which concepts for measuring profitability of dialog marketing affine companies are applied—in % (multiple answers possible, n = 70) (*Source* Author illustration, data source: Kreutzer & Schober, 2010)

the awareness and the image of companies and brands, the readiness to purchase, and—somewhat analytically more demanding—the brand value. However, the extent to which various marketing activities have contributed towards which sales or preferably profits still remains unconsidered. The fact that the **cognitive firewall between marketing and sales** that can still be frequently found in companies needs to be taken down for such findings goes without saying. The most convincing way of overcoming such a firewall is the **obligation to achieve mutual, identical objectives.** If necessary, the boards responsible will have to bite the bullet when marketing and sales have to be accounted for in the various board departments!

> **Think Box**
> - According to which performance indicators can marketing be measured in my company?
> - Is there a focus on potential-oriented objectives when evaluating marketing activities?
> - Or is the extent that marketing contributes towards achieving sales and profit targets monitored?
> - How high is the cognitive firewall between marketing and sales? Do they encounter each other as "inhabitants of different planets" or do they work together with regard to identical targets?
> - Who in my company can give decisive impetus to call for the focus on results in marketing?

Talks with many of the companies we look after have repeatedly shown that the following **questions** still remain far too frequently **unanswered** in companies:

- Who are my "best" customers and what is used to measure the "best" (turnover/contribution margin—as past/current value or as a forecast)?
- How loyal are the customers and what is used to measure "loyalty" (length of customer relationship, amount of turnover—absolute or relative as in share of wallet, extent of successful recommendations)?
- What segments do customer retention measures focus on today—and why?
- Which acquisition channels and measures win the best/worst customers—and why?
- Which offers win the best/worst customers—and why?
- Which support services help retain customers most efficiently—and why?

Without answering the abovementioned questions in detail, the company is off in **blind flight mode**. Acquisition concepts are continued even though it is not at all clear whether these can help win value-creating customers in the long run. Customer support is focused on certain customers—those with the largest turnover or the greatest (board) presence as the case may be—irrespective of whether these customers actually generate the greatest value contributions for the company. Measures for customer retention are drawn up without checking in detail which customers—with regard to their future results potential—are particularly valuable for the company. It is essential to overcome such situations.

> **Think Box**
> - What is the situation in my company—are we also off in blind flight mode concerning the impact of marketing and sales activities?
> - What significance is given to customer value in my company?
> - Are measures due to the customer value controlled in my company?
> - Where is the responsibility for the implementation of a value-oriented customer management system to be placed?

If such questions are not answered or not answered in a substantiated way, **marketing management** shall neither reach its **efficiency goals** ("doing the right things") nor its **effectiveness goals** ("doing things right"), because tough monetary evaluation criteria are lacking. Then, if anything "We don't actually know where we want to go, but we are getting there!" The necessity of overcoming this situation arises from the fact that the CMOs are facing a greater challenge than in the past to consider their responsibility for the ROI. If **marketing** is to position itself not only as a **strategy leader** but also as a **profit driver** within the company, marketing activities must be directed towards the **confirmability of the profitability of one's own doing**.

This results in the necessity of applying a feasible evaluation concept that helps to answer the abovementioned questions. The qualified **determination of the customer value** thus represents the **basis for a value-oriented customer management system**. This means the development and implementation of concepts that

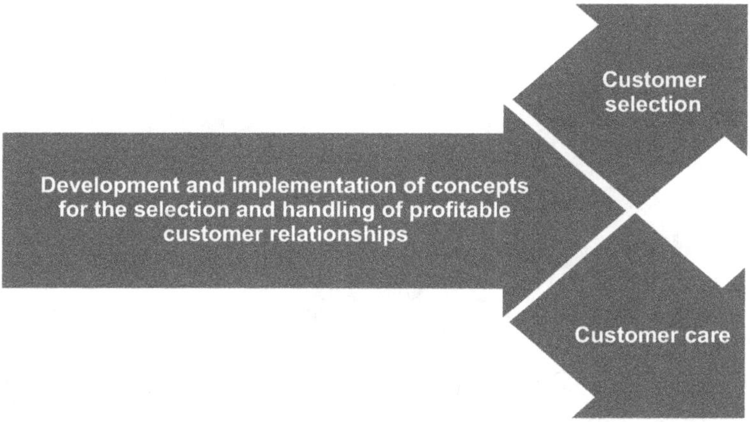

Fig. 5.3 Directions of value-oriented customer relationships management (*Source* Author illustration)

contribute towards the **selection and processing of profitable customer relationships** (cf. Fig. 5.3). It becomes obvious that value-orientated customer management involves two key tasks: on the one hand, the **selection of customers to be won or retained,** on the other hand, the specific **organization of customer care**.

In summary, the **tasks of a value-oriented customer management system** can be called selection, setup, design, retention, and termination of business relationships with individual customers or customer groups on the basis of their value contributions towards defined company objectives (cf. Helm & Günter, 2006, p. 11).

Nowadays, if one analyzes the way the value contribution of individual customers is operationalized—if at all—then the results in Fig. 5.4 are frequently seen. Apart from companies that do not or cannot classify their customers at all, there are rather general descriptions such as **good/bad customers** whereby it still remains unclear what is exactly meant by that. **Major and minor customers** are another classification behind which the expectation that can frequently be refuted in some cases lies hidden that a major customer is automatically also a profitable customer. Even the differentiation between **casual customers and regular customers** or between **online and offline customers** do indeed represent a behavior pattern of the customers, yet say nothing about the levels of turnover and contributions for the company.

The **ABC analysis** does help with this somewhat because at least the distribution of the customers depending on their level of **turnover** or **contribution margins** achieved is represented. The concentration effects becoming visible can provide significant orientation for the focusing on "management attention" as well as retention measures. However, in the classical ABC analysis, "behavior from the past" is rewarded in the "here and now," whereas future high potentials still to be found in segment B or C today have to do without value-oriented support. For their

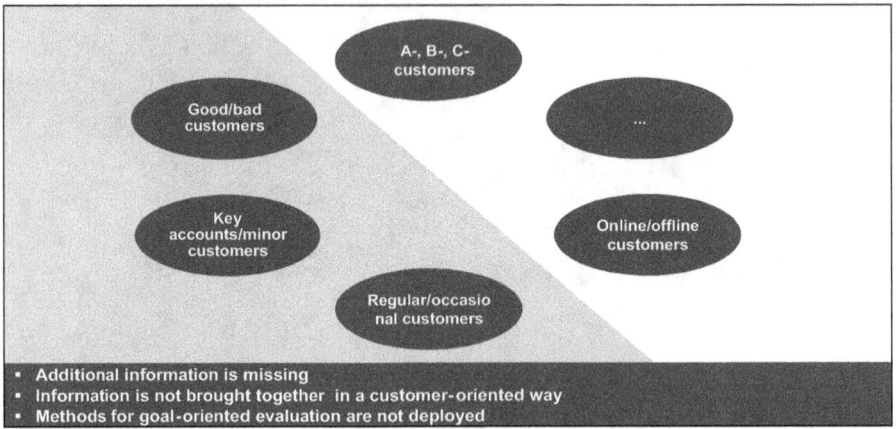

Fig. 5.4 Realization of the customer value approach in many companies—nowadays (*Source* Author illustration)

potential has not yet been considered. It is thus essential to create a greater information density in order to exploit the awareness potential for value-oriented customer management.

The **dominating sources of error**, which we need to eradicate when setting up a value-oriented customer management system, are shown in Fig. 5.5. First of all, an **ex post assessment of customer value** is conducted without critically questioning whether the behavior displayed by customers in the past can also be expected in the future. If I call the buyer of a new car a top customer today, then I neglect the fact that he probably won't be buying another one for the next 2 or 3 years. Such an approach thus systematically prevents the customers with potential for development (the "buyer of a new car—tomorrow") being recognized and thus processed in an appropriate manner.

A further area of concern is **static models** that rely on mere trend extrapolation according to the slogan "more of the same," without anticipating possible breaks in the system and considering them when determining customer value (e.g., a customer moving from an offline to an online channel). This is where the informational access to the world's largest preferences database—*Facebook*—can provide very interesting insights (cf. Chap. 7).

An **undifferentiated approach** is then existent if, during the evaluation of customer values, it is not taken into consideration that diverse customer groups can develop in different ways with time. In addition, the fact of frequently ascertained **customer reviews being independent of campaigns** must be criticized. At the same time, it remains unconsidered that there can be a great difference regarding the customer value depending on the way a customer is approached or which offer is supposed to be submitted. A customer can be predestined for being approached by telephone and thus represent a high customer value for a sales campaign. Yet, as the same customer almost never responds to

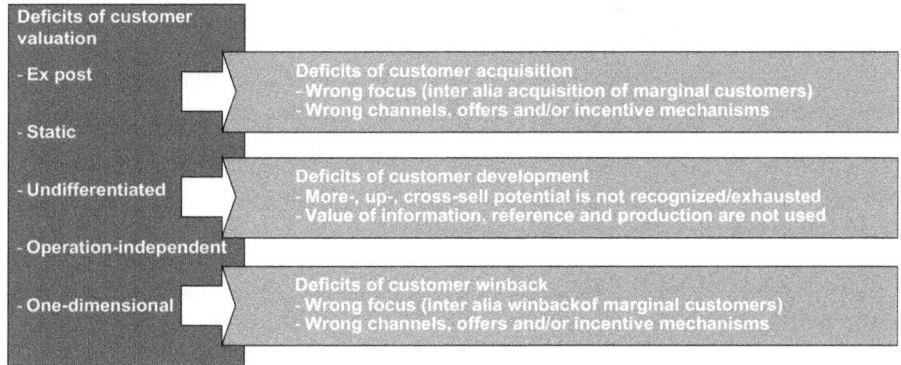

Fig. 5.5 Sources of error when controlling customers (*Source* Author illustration, data source: Helm & Günter, 2006, p. 24)

being approached by email, his value for an email approach will turn out to be correspondingly low. With a **single dimensionality of customer value evaluation,** merely one criterion is drawn to determine the value. This is frequently the turnover without considering the fact that this does not positively correlate to the contribution margin in all customer groups.

The consequences of such an approach are **deficits in customer acquisition** (cf. Fig. 5.5). By means of an insufficient customer value evaluation, marginal customers, i.e., customers merely relevant for the company "on the edge," who produce a negative contribution margin are won over. In addition, communication channels, offers for winning new customers, and/or specific incentive mechanisms are applied which do not lead to long-term valuable customers. As well as that, **deficits in customer development** emerge because more-, up-, and cross-sell potential is not recognized or cannot be exploited in a suitable way. Furthermore, a possible information, reference, and production value of customers is not used because it is simply ignored during the evaluation. Thus, a company can repeatedly bet "on the wrong horse" because the relevant control information is missing. Last but not least, **deficits in winning back customers** also materialize because the focus is on the wrong place. As the case may be, "marginal customers," who are only interesting for the company on the edge, can also be won back. Or, on the other hand, wrong channels, offers, and/or incentive mechanisms are applied (cf. Fig. 5.5).

> **Think Box**
> - How is the customer value calculated in my company—if at all?
> - How frequently is the underlying concept of evaluating customer value examined with regard to its power of realization?

(continued)

- Who is responsible for the anchoring of customer value in the course of customer acquisition, customer development, and customer reacquisition?
- Are employees assessed and rewarded on the basis of the achieved customer value?

In contrast, a customer management system consequently based on a meaningful customer value can **prevent a company's campaign-related blind flight** and set suitable focuses during the acquisition and support of customers (cf. Fig. 5.6). Then the right focuses can be set in the **sphere of acquisition** in approaching and channels, with initial offers as well as with cooperation work for customer acquisition to be entered into. In the **sphere of care,** customers with potential for development can be focused on and the suitable customers can be integrated in retention programs. At the same time, it becomes clear, which customers should be won back and which one is happy to say farewell to.

Figure 5.7 shows which concepts and criteria can be applied when evaluating the customer value. First of all, when it comes to the **time reference**, one can differentiate between whether—as described above—a mere **ex post consideration** takes place and whether attempts are made to forecast future developments (**ex ante approach**). Many companies still prefer the ex-post consideration even though it is not really **relevant for control**. In addition, it must be checked whether a **one-period approach** (e.g., be limiting it to a 6-month ordering season in online trade or an entire business year) should be conducted or whether several periods should flow into the evaluation of customer value. The respective decision depends on the seasonal deviations a field of business is subject to. In online fashion trade, it may, for example, be wise to assess the spring/summer and autumn/winter seasons separately, because there are frequently mere seasonal buyers. A **cross-period**

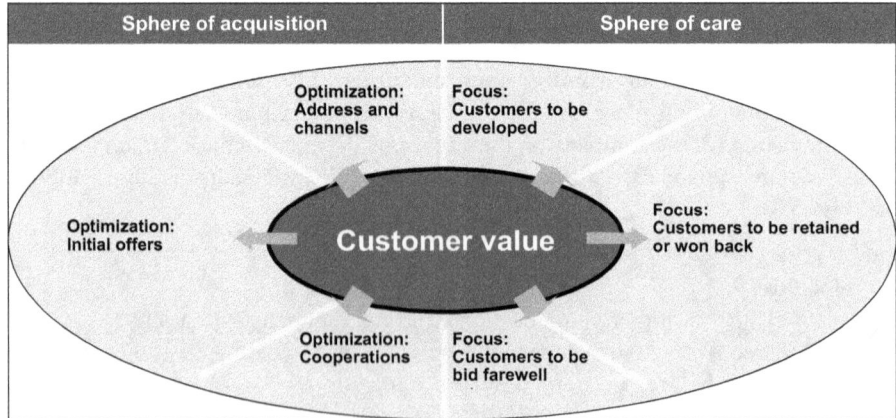

Fig. 5.6 Meaningful customer values for focusing marketing activities (*Source* Author illustration)

Concept	Development/criterion
Time referece	• Ex post vs. ex ante • Single vs. multiple period consideration
Unit	• Individual customer • Customer groups
Temporal modelling	• Static approach • Dynamic approach
Content modelling	• Single-vs. multidimensional concepts • Monetary vs. non-monetary criteria
Value realisation	• Nominal value consideration • Discounting of value to time of analysis
Driver of customer value	• Turnover (more-, cross-, up-sell-oriented) • Contribution margin • Reference value (image effect of customer, opinion leader or multiplier role, recommendation value of customer) • Information value (customer as brand ambassador, as creative partner) • Production value (customer as co-producer) • Transaction costs (customer driven care costs) • Transaction costs (company driven care costs)

Fig. 5.7 Concepts and criteria of determining customer value (*Source* Author illustration)

consideration would otherwise make an average customer out of a "top customer spring/summer" and a "noncustomer autumn/winter" with regard to the customer value—and would not do justice to the customer specifics.

With the **unit of consideration**, a differentiation must be made as to whether the "segment of one," as in individuals or companies, is to be evaluated or whether various **customer groups** are considered. In principle, a **single customer-oriented consideration** is to be preferred with efficient analysis and prognosis methods, because the individuality of every customer can be taken into consideration. Often only an individual customer assessment can supply the necessary information for individual customer care.

With **temporal modeling**, there are **static concepts** at first, which contain an extension of the time course observed in the past in the future (extrapolation). In contrast, **dynamic models** attempt to consider further factors of influence in the prognosis and can thus principally achieve a higher quality of the prognosis.

With respect to **content modeling**, the number of dimensions included must first be differentiated. **Single-dimension approaches** frequently focus on turnover or contribution margin. **Multidimensional models** in contrast attempt to take the various facets of a customer value into consideration and to thus do justice to the differentiating level of value of the customers, beyond turnover and contribution margin. At the same time, with content modeling, a differentiation must be made between the approaches that consider **monetary factors** (such as turnover) and **nonmonetary factors** (such as the reference value of a customer).

The **realization of value** can abstract factors determining customer value from the "pay-in or pay-out point in time" and **consider the value in pure nominal terms**. This leads to turnover generated with the customer in 2 years having the same level of value as turnover achieved in the month to come. The concepts that discount the future **ingoing and outgoing payments** on the time of the evaluation must be differentiated from this. In doing so, the underlying rate of interest during discounting is given a significant meaning as it decisively defines the "cash value" to be determined.

The large bandwidth involved is with the **drivers of the customer value**. In this case, a differentiation must be made as to which performance indicators and which cost categories are taken into consideration when defining the customer value. The most frequent criterion is still **turnover**, whereby, with regard to the future development of turnover, the more-, cross-, and up-sell potentials are often not differentiated in the long term. In many cases, only the previous turnover with regard to the more-sell potential is taken as the basis. An important yet not universally applied variable of success is the **contribution margin**, which needs to be defined here in terms of customer-specific or customer group-specific as the sum of the contribution margins of all products purchased by them or services used by them. Many companies still fail in determining such contribution margins by the customer. Figure 5.8 illustrates the insights that can accompany the focus on contribution margin. Based on a simple **customer group analysis according to contribution margin,** with an own company project, the following meaningful results become obvious for a finance services provider:

- With 40 % of the customers (so-called D-customers), only 5 % of the result is generated.
- 10 % of the customers (A-customers) contribute 50 % to the profit.
- An in-depth analysis reveals in addition that 66 % of the D-customers reveal a negative contribution margin and 21 % of the D-customers only generate a slightly positive contribution margin.

The relevance of such an approach almost imperatively results from the **Pareto principle** already mentioned. When evaluating the customer base, it is not only

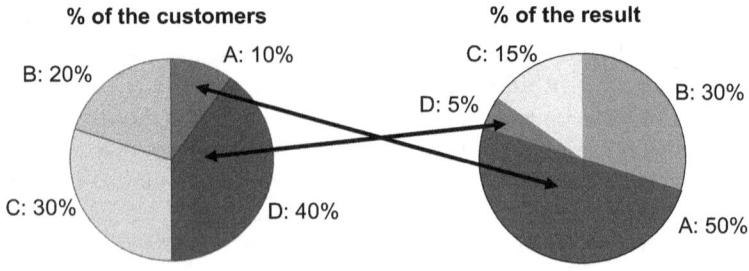

Fig. 5.8 Monetary consideration of results per customer group (*Source* Author illustration)

revealed that 20 % of the customers frequently cause 80 % of the problems. It is more important that 20 % of the customers can also stand for 80 % of the turnover or, better still, the contribution margin. If this calculation is pursued, then 20 % of the 20 % of the customers (i.e., 4 %) stand for 80 % of the 80 % of the contribution margin (i.e., 64 %). When continued, this possibly even means that 20 % of the 4 % of the customers (i.e., 0.8 %) stand for 80 % of the 64 % of the contribution margin (i.e., 51.2 %). Taking this to the extremes, the consequence is that 20 % of the 0.8 % of the customers (i.e., 0.16 %) stand for 80 % of the 51.2 % of the contribution margin (i.e., 40.96 %). If this insight is translated to a football stadium that is filled with 50,000 own customers, 80 of the customers present here stand for 41 % of the contribution margin! One merely needs to recognize these 80 customers in their entirety (cf. Peppers, 2012). The astonishing thing about this is that this knowledge of the **intrinsic value of one's own customers** must remain hidden from competitors—however, ideally not hidden from our own company and our own employees!

Without this in-depth **transparency of the value creation with one's own customers,** customer acquisition and care cannot be optimized. An unchanged continuation of previous measures would presumably create or maintain similar customer structures. With that, decisive potentials for increasing company revenue in the long run remain unconsidered. As a result, "**value-driving**" and "**value-destroying**" customers must be identified at an early stage.

> **Think Box**
> - How could an efficient customer value model look for my company?
> - Which factors need to be incorporated?
> - Can the customer value as a target indicator be connected with incentive structures of the employees responsible for marketing and sales?
> - Who could be assigned the task of the development and ongoing implementation of a value-oriented customer management in my company?

An **analysis of the customer bases** can easily provide very interesting information if the attributes **loyalty** (duration of customer relationship) and **potential exploitation** (number of products or services drawn on) are taken into consideration. Figure 5.9 illustrates such an analysis for an online retailer. A **yo-yo customer** is the type of customer who repeatedly buys individual products offered by the retailer (e.g., shoes) but does not build up loyalty. In the case of such a customer, attempts must be made to offer him other products in order to commit him to the company in the long run. A **product focuser** is loyal to his company. However, the customer relationship is restricted to one product or a few articles. It is imperative to give concrete cross-sell impulses (e.g., with regard to clothing offers) in order to exhaust the customer's potential. If there is successful access to their *Facebook* data, then targeted and thus highly relevant offers can be submitted. And, of course, there must be extensive monitoring as to whether the measures taken were also

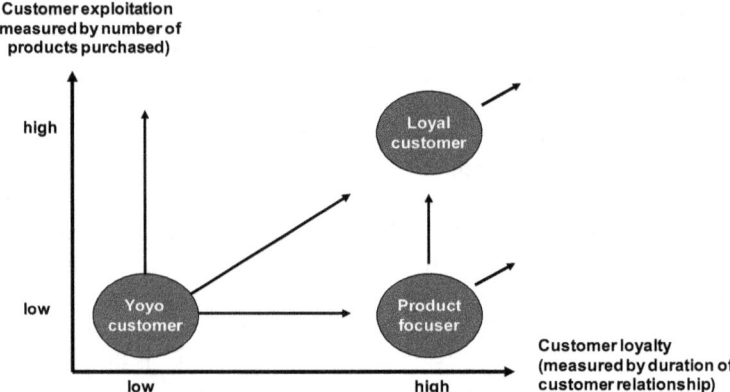

Fig. 5.9 Approaches for increasing and safeguarding return at customer level (*Source* Author illustration)

successful (cf. Chap. 7). The **loyal customer** with numerous product purchases, in contrast, must always be provided with good arguments as to why he should remain loyal to his provider. Without an **analysis on the basis of individual customers**, no differentiated impulses can be developed. And once again, it is essential to—based on the customer value—determine whether a further differentiation of customer care is also "rewarded" by higher customer values. For there must not be an individualization of customer care for its own sake!

> **Think Box**
> - How systematically do we approach customers—based on the customers' previous behavior patterns?
> - Have we ever analyzed such "typical" behavior patterns of our customers?
> - Do we avail ourselves of various support concepts for different customer segments?
> - To what extent do we review whether this form of individualization is worth it?
> - Who is responsible for the development and implementation of it?

It goes without saying that **winning back lost or inactive customers** is a continuous challenge. The question however is: Do we actually notice when customers opt out of the relationship, even if they don't need to give notice? Do we notice, for example, that a customer hasn't made a purchase for 2 months although he usually visits our online shop every month and also places an order there? Or does such a regular analysis not take place?

Which contributions to revenue are to be targeted by a consequent **analysis and activation of (inactive) customers** becomes obvious based on the following calculation. Should, for example, a company succeed in retaining only 20 % of the 10,000 customers leaving per year, who generate an average contribution margin of US $150, this involves **safeguarding revenue** of US $300,000 per year. If, by means of appropriate online measures, 5 % of the total number of customers of 100,000 can be motivated to purchase an additional product with an average contribution margin of US $45, an **increase in revenue** of US $225,000 can be achieved (less the respective campaign costs). This makes it evident which **lever for increasing revenue** goes along with a founded customer analysis.

> **Think Box**
> - How well developed is our customer retrieval management system?
> - Do we actually notice when customers become inactive?
> - Do we regularly register which customers have become "inactive" because they have broken out of their personal sales behavior?
> - Are there appropriate analysis processes for this?
> - Are there any action plans that are used for a systematic reactivation—to be initiated automated or merely manually?
> - Where can the corresponding responsibility be assigned to?

To the extent that contribution margins are determined on the side of the customer, only the offer-related costs are usually incorporated. In such a determination of contribution margin, the **transaction costs** involved in a customer relationship are less frequently considered. These include first of all the **customer-driven care costs** the customer causes by his specific behavior pattern (by frequently returning products, many calls to the customer service center, slow payment causing interest and processing costs). In addition, the **company-driven support costs** must be calculated. These include, for example, costs depending on the number and contents of advertising campaigns (e.g., sending out high-quality catalogs), approaching customers by phone or email, as well as visits by field workers.

Figure 5.10 illustrates an example of a **customer value model in an online shop**, which contains the customer-driven care costs in the shape of a scoring model. For the majority of the companies, such an extensive compilation of the transaction costs represents a challenge for the future.

In order to develop such a **scoring model**, it must first be determined which **criteria** should be used to evaluate customers. Then these criteria must be given a **weighting**, in order to express the difference in significance of the various criteria. The weighting factors must add up to 1.0. In the course of this determination, intensive discussions frequently take place in companies because, when developing such a scoring model, rather intuitively formed evaluation patterns become transparent and thus also debatable. This alone embodies a great value of this approach. Care must be taken that the criteria are as independent from each other as possible, in order to avoid an undesirable multiple consideration of the same data.

Criterion		Score					Score x Weighting
	Weighting	1	2	3	4	5	
Average service intensity per month	0.3	Call > 0.5	Letter/fax > 0.5 or Call <= 0.5	E-mail >= 0.5 or Letter/fax <= 0.5 and call = 0	E-mail < 0.5 Other 0	0	
Turnover per month in €	0.5	x < 10	10 >= x < 25	25 >= x < 50	50 >= x < 100	x >= 100	
'Tell a friend' advertising (number of new customers per month)	0.2	0	x < 0.5	0,5 >= x < 1	1 <= x < 2	x >= 2	
Total	1.0						Sum

Fig. 5.10 Scoring model for determining customer values in an online shop (*Source* Author illustration)

In the next step, all **criteria must be operationalized**, i.e., made measurable and be **given points (scores)** with regard to their varying developments. The multiplication of the points given with the respective weighting totaled across all criteria results in a total score for each customer. This facilitates a comparison of the customers based on their individual contribution to the company.

The advantage of such scoring models is—as shown in Fig. 5.10—that qualitative and quantitative criteria can be integrated in the analysis. Apart from that, subjective evaluations (the famous "gut feeling") are compacting to a total evaluation by the involvement of several people. The documentation of the evaluation mechanism enables a review of how true the evaluations of the customer values made were after a year, for example. By doing this, important requirements are created for a "learning organization." For experience with the scoring approach can be used to optimize other processes.

The **intrinsic value of a customer**, however, is not restricted to these monetarily comprehensible factors. Thus a customer—in the B2B as well as in the B2C market—can also have an important **reference value** for the company (cf. Fig. 5.10). If a famous personality or a well-known company (e.g., *Google* or *BMW*) uses its own product, there is thus a significant **image effect** that needs evaluating. The same applies to an **opinion leader or multiplier role** that a customer can take on and which, for example, is reflected in the number of customers acquired by "telling a friend" or by positive posts or tweets via *Twitter* and leads to the **recommendation value of the customer**. In addition, a customer can also be evaluated regarding his **information value**, if he becomes active for the company as a brand ambassador or creative partner. A more intensive relationship is expressed by the **production value**, if the customer becomes a codeveloper or coproducer for a company. People who provide exciting user-generated contents in the course of crowdsourcing processes can achieve a high production value for the company in this way—even if they are no actual customers. However, we are well advised to also determine their value contribution for our own company.

The overall objective should be to determine the **customer lifetime value** (CLV) for every customer. This represents the sum of the value contributions of a customer defined in line with one or several of the criteria described above. In doing so, the value contributions are aggregated over the possible or aspired duration of the relationship with the company. The values achieved facilitate the decisions about which investments can be made in the long-term retention of a customer. With this in mind, besides the **human being**, we should motivate ourselves and our employees to also see every person coming through an online or offline door as the **value bearers** who—when looked after well—may generate US $2,000 or 20,000 or even 200,000 turnover or contribution margin for the company over the years to come. Some concepts go to the extent of recommending that the employees imagine a kind of "value image" on the forehead of the potential customer!

We are well advised to elaborate a specific **concept for determining customer values** with our industry specifics, our own targets, as well the availability of relevant data in our company in mind. We should then aspire a greater incisiveness of determining customer value if we have the possibility of carrying out a differentiated care of the customers geared towards it. The **diversity of customer evaluation** must thus go hand in hand with the **diversity of approaching customers**. The starting points for a company-specific approach are summarized in Fig. 5.11 as **competence pyramid for determining customer values**.

According to our experience, **setting up of such a competence pyramid** is frequently a project taking several years. The starting points for an underlying customer evaluation must be explored based on the possibilities of differentiated customer care. When **implementing a value-oriented customer management system**, one must, among others—even and especially in marketing—say farewell to much-loved habits when looking after customers:

- **No customer orientation at any price!**

  Customer orientation is no end in itself but a means to an end. Customer orientation must make a contribution towards achieving superior marketing and company targets—no more, no less.

- **No aspiring maximum customer satisfaction!**

  Every investment in increasing customer satisfaction must be worthwhile for the company.

- **No desire to retain all customers!**

  A customer value evaluation is to find out which customer relationships a company is particularly interested in or should be.

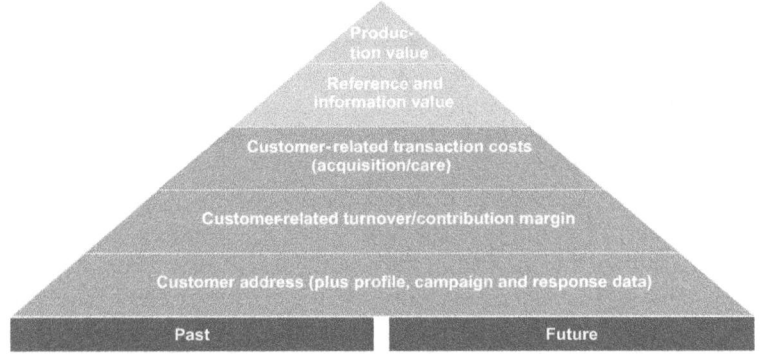

Fig. 5.11 Competence steps for determining customer value (*Source* Author illustration)

- **No equal treatment of all customers!**
  Customers that generate more added value for a company may and must be worth more to a company in terms of support.

  > **Remember Box** Customer acquisition and customer care are to be developed in a value-oriented way and to be brought in line with the sales orientation!

In a slightly modified version of a quote by *Peter F. Drucker*, it should thus say: "The objective of a company is to create a **p r o f i t a b l e** customer." Merely creating a customer is no longer enough!

> **Think Box**
> - How high is the customer lifetime value in my company?
> - Are there precise calculations and starting points about the extent we succeed in achieving this value?
> - Do approaches already be recognized in my company for setting up a competence pyramid for a customer value evaluation?
> - How relevant is such a concept for us?
> - Which levels should be aspired?
> - To what extent are "customer orientation at all costs," "maximum customer satisfaction," "a desire to retain all customers," as well as "equal treatment for all"—still—aspired at my company?
> - Under whose responsibility does it fall to overcome these wrong ideals?

We all know that customers are the only performers who permanently "put money into the company"—if we, as providers, do a good job. That is why we have to focus all of our activities—not only those in marketing—on the target of generating customer value. The fact that this flow of income also represents the requirement to take on social responsibilities with financial support must be mentioned at this point. In sum, only those companies that also manage to generate "customer value" by means of "value for the customer" in the long run will survive digital Darwinism. For this reason, **all changes in our business model** as well as the **engagement in social media** must be based on their **contribution towards achieving customer value**. For it is still imperative that:

> Turnover is the applause for the company!

And it is our responsibility as a manager to turn turnover into profitability for the company by extensively "controlling" all activities and recognizing the **customer value as the "key" KPI**.

Quick Wins

- _____
- _____
- _____
- _____
- _____
- _____
- _____
- _____
- _____
- _____
- _____
- _____
- _____
- _____
- _____
- _____
- _____
- _____
- _____
- _____
- _____
- _____
- _____
- _____
- _____
- _____
- _____
- _____
- _____

Confidence: The Latest Currency in Marketing and Management   6

In Chap. 5, we dealt in detail with how marketing can become the ROI driver in a company. In the course of this, the **do ut des principle** was addressed. However, it is not only found—on a high degree of abstraction—in terms of the balance between the "value for the company" and "customer value." It also has a long-term influence on many concrete measures of the company itself if it, for example, involves the provision of information for the company as a requirement for individual customer care. It is imperative that potential customers and customers have to provide "additional information" in order to be able to experience "added individual support." This important insight has still not been accepted neither in the broad segments of politics nor by the "consumer protectors" nor by many customers themselves. The substance of this insight culminates in the **law of disproportionality of information** (Kreutzer, 2009, p. 69):

> The more information available about a consumer or a decision-maker or a company, the more accurate offers can be placed. This means that we require more information about potential customers and customers in order to provide them with less yet relevant information.

The **task of communicating**, which we, as a company, still have to fulfill with regard to the abovementioned target groups, can be found in Fig. 6.1. Due to the hefty headwind that is confronting us with this topic from the "consumer protection" corner, we are well advised to make use of the **cooperation potential** between the companies as well as the respective organizations as described in Chap. 3, even when implementing this plan.

The implementation of the task involved for companies comprises an internal and an external challenge. The **internal challenge** comprises **merging** and **consolidating in a central place** the **multitude of information** about potential customers and customers emerging in the various parts of a company. Thus, in the controlling department, there is information about paying behavior; in the customer service center as well as the Internet division, information about inquiries, orders, and complaints; and in a shipper's returns center, data about returning behavior.

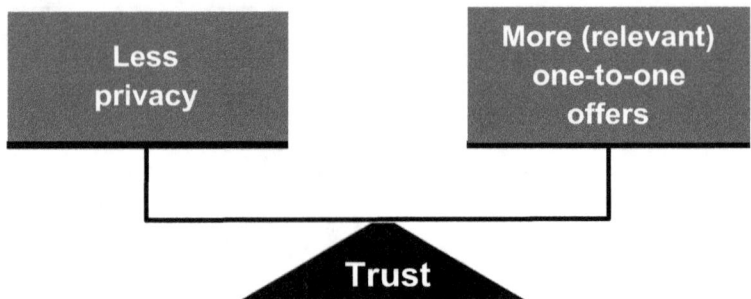

Fig. 6.1 Confidence as the basis for the balance of "data contra one-to-one offers" (*Source* Author illustration)

This information needs to be made available for all of those working in a customer-oriented way. This can be a call center agent, the data analyst, or someone responsible for communication or sales. This all involves IT systems and IT processes that lie in the hands of the companies to a large extent.

The **external challenge** comprises **gathering** as much **information about customers** as possible. And the prerequisite for obtaining more information from them is **confidence**! It is not until we manage to establish confidence with the relevant target groups for our actions that they will also grant us access to sensitive data necessary for efficient (social) CRM. This aims at **establishing a learning relationship** (cf. Fig. 6.2). This is a relationship in which we as a company "learn" more and more about our customers and their preferences step by step—in order to then attend to them in a more customized way (cf. Peppers & Rogers, 2011, p. 1).

Confidence is an **important and valuable currency** in equal measure. However, it cannot be bought—it can only be earned! For **confidence is like an account**: you first have to pay into it for a long time before you can withdraw significant amounts! Confidence is the prerequisite for potential customers and customers filling out our questionnaires, for receiving their permission to contact them by email or telephone—and customers giving us the **token** as access to *Facebook* or *Twitter* data services like PeerIndex use this kind of access to get additional data (in detail in Chap. 7).

Such a token opens up the data flow necessary for an **efficient CRM** which covers, among others, the data displayed in Fig. 6.3 (cf. Chap. 7 for more detail).

Even if we have already pointed out the insight that **"public is the new private,"** the following still applies: whether we as a company are allowed to access the "socially spread" information depends on the earned **confidence potential**. And confidence is also the prerequisite for potential customers and customers (positively) engaging in our company, our brands, and offers and being creatively active for us, for example, in the category "earned media." The use of social log-ins is also based on confidence in how the "taking company" handles my data. And only with

Learning Relationship

Fig. 6.2 Establishing a learning relationship (*Source* Author illustration)

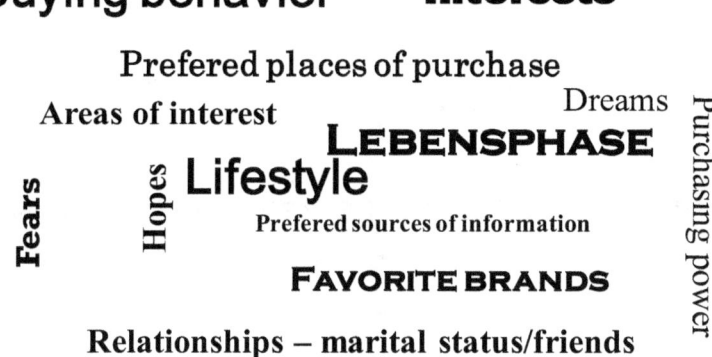

Fig. 6.3 Which data about our potential customers and customers we need (*Source* Author illustration)

a basis of confidence can the corporate posts at *Facebook*, which frequently generates the highest number of responses, also work: "How was your weekend?"

Due to the relevance of confidence, here we can—in addition to the three-dimensional CRM in Chap. 1—integrate confidence as in "emotional proximity" as fourth dimension. The **concept of four-dimensional CRM** can be found in Fig. 6.4.

In **classical business models**, which were still geared towards the (physical) proximity of store and personnel, the old form of the currency "confidence" was defined via the **personal familiarity** in the store, the one-to-one advice, and the personal recommendations. In the **digitalized business models,** other indicators are required to determine the status of confidence in a company. Here, the personal contact from person to person is frequently lacking, being replaced by the relationship between person and company. With that, the **company as an anchor of confidence** is given very high significance because—in the eyes of the customers—it increasingly replaces the human point of contact. In order to determine the **status of confidence of potential customers and customers in one's own company**, there are various concepts and KPIs. Among the most important **confidence KPIs** are, among others, the following indicators:

Percentage of customers who
- Have given permission for email contact
- Have given permission for telephone contact
- Have given a token for accessing *Facebook* data

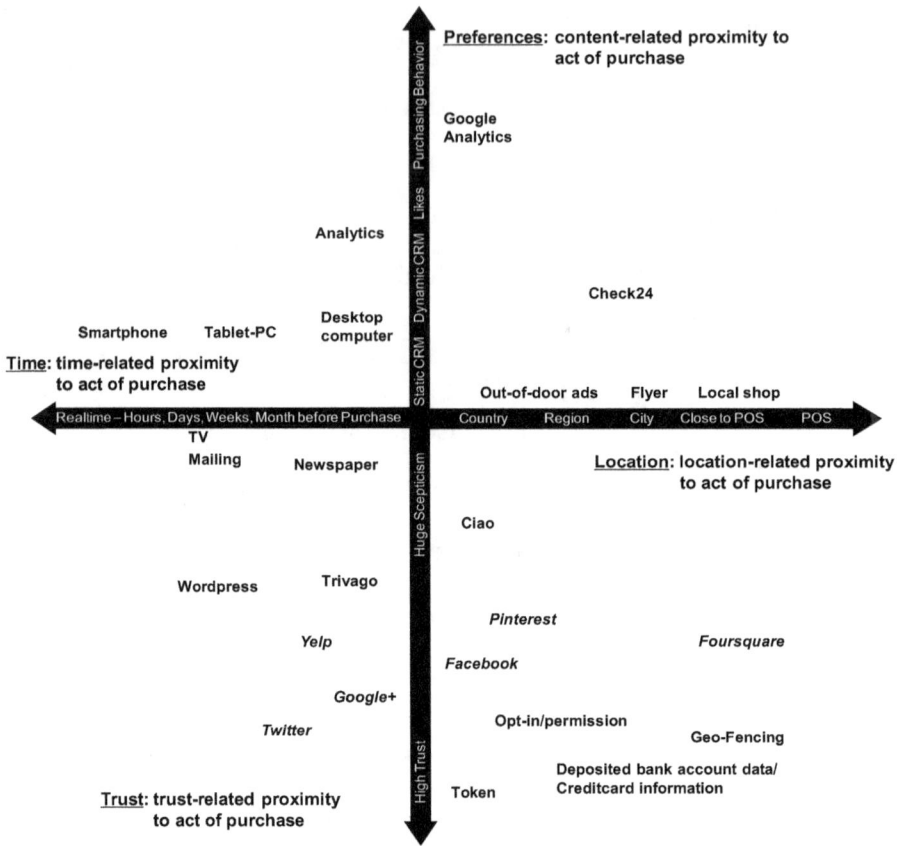

Fig. 6.4 The concept of four-dimensional CRM (*Source* Author illustration)

- Have provided bank or credit card access data for direct debiting
- Follow us on *Twitter*
- Have liked us on *Facebook* or *Google+*
- Have pinned us on *Pinterest*
- Have (successfully) recommended an offer to others
- Have given permission for geo-fencing

Geo-fencing is a mixture of geography and fence and describes a virtual border. It facilitates the **localization of objects**. If a customer gives permission for geo-fencing, then the company can identify this person as soon as they, for example, enter the radius of their own stationary retail store. GPS and RFID form the basis for this. It makes it possible to provide personalized and customized messages by mobile when the customer is within close proximity. The requirement for corresponding permission in this case, too, is that the customers have begun to trust the company. This is where the **power of four-dimensional CRM** is displayed.

We are well advised to systematically gather **confidence KPIs** and see them as **confidence indicators**. If we succeed in getting a glance at the situation of the main competitors, **confidence benchmarking** can be performed. Even if we do find it difficult at the beginning to recognize the significance of the **"confidence currency"** for our entire company—we should still set out on this journey!

Think Box
- What significance have we given to "confidence" in relationships to potential customers and customers?
- Have we ever measured the intensity of confidence?
- What is the status of confidence of our customers with regard to the named confidence KPIs?
- Have these indicators ever been gathered or analyzed?
- Were they interpreted as evidence of the confidence gained?
- What is my company's standing in comparison to the main competitors?
- In whose field of responsibility does the "monitoring of confidence" belong?

Mom and Pop Store 3.0

How a **stationary store concept built up on confidence** can look is demonstrated here with the example of the *PayPal Futurestores,* which was presented as a **concept study** based on the fictional *Hudson + Vestry shops, New York*: *Sharon* is on her way to work at 8 in the morning. She walks past her favorite boutique and discovers a great handbag in the window. As the boutique is still closed, she takes out her smart phone, activates the *RedLaser* app, and scans the "object of desire." As *Sharon* has given **permission for geo-fencing,** the stationary store can now communicate with her. Not only does she receive the price information but also the message that there is only one more of these handbags left in the store. *Sharon* activates the function "please put aside until this evening" and relieved; she goes to the office.

At 7 p.m., *Sharon* finishes work for the day. The minute she enters the **geo-fencing radius** of the store, a message is sent to the store manager. Not only does he see that *Sharon* is coming closer to the store, he is also shown that *Sharon* invests a significant share of her income in his store every month. Her profile displays three stars, labeling her a key customer (**customer value**)! As her recent purchases as well her preferred brand are stored in the system—including clothes and shoe size—based on the "please put aside," the store manager was able to prepare an **individualized "presentation table"** for *Sharon* with further suitable products.

A little later, three customers entered the store including *Sharon*. All of them are greeted in an appreciative manner but the store manager, working alone in the store today, only approaches *Sharon*. The two other customers are—according to the database—known as "lookers," who like looking around but have never bought anything. On the prepared presentation table, *Sharon* finds not only the handbag she had chosen but also a pair of shoes (in the right size) and a scarf matching in color. She cannot resist either. She pays by *PayPal*—upon request even in installments. At the same time, a voucher for US $10 that *Sharon* had received by mail 10 days earlier is automatically deducted. *Sharon* leaves the store with a very good feeling—and will happily come back again!

This is how a **business model based on confidence of the future** can look—and we won't have to wait for this future much longer. All of the necessary technology is available. And as soon as customers notice that using it means a great extent of convenience for them, it will become more accepted—even for the granting of the necessary permissions. And *PayPal* has good reasons for working on such an **ecosystem for stationary retail stores**, in order to promote its payment model in the long term.

The **Net Promoter Score (NPS)** is an efficient and equally easily applicable concept to determine the **extent of the emotional binding and confidence.** Essentially, determining the NPS is about the question as to how many of a company's customers recommend the company (net) to others. The basic concept of the NPS is described in Fig. 6.5.

In order to **determine the Net Promoter Score**, one single question is asked: "How likely are you to recommend our company/product/service to your friends and colleagues?" The answers can be given on a scale from "0" (not at all likely) to "10" (extremely likely). **Promoters** of a company or a brand are those people who give a score of "9" or "10." **Detractors** are those people who, with regard to recommending it to others, only give a score between "0" and "6." **Passives** are those people who give a score of "7" or "8." When working out the net value of those recommending, the percentage of detractors is subtracted from the percentage of promoters. The group of the indifferent is not taken into consideration. Thus the **calculation formula of the NPS** is:

▶ NPS = promoters (in %) – detractors (in %)

The **values of the NPS** can be "100 %" in the best case if all customers have given the score "9" or "10." In the worst case, the result can be "−100 %" if all customers have only given scores between "0" and "6" (cf. Reichheld, 2003). Even if the significance of the NPS has frequently been questioned critically (cf., e.g., Keiningham, Aksoy, Cooil, & Andreassen, 2008), it is recommended for use in companies.

Question: How likely are you to recommend our company/product/service to your friends and colleagues?

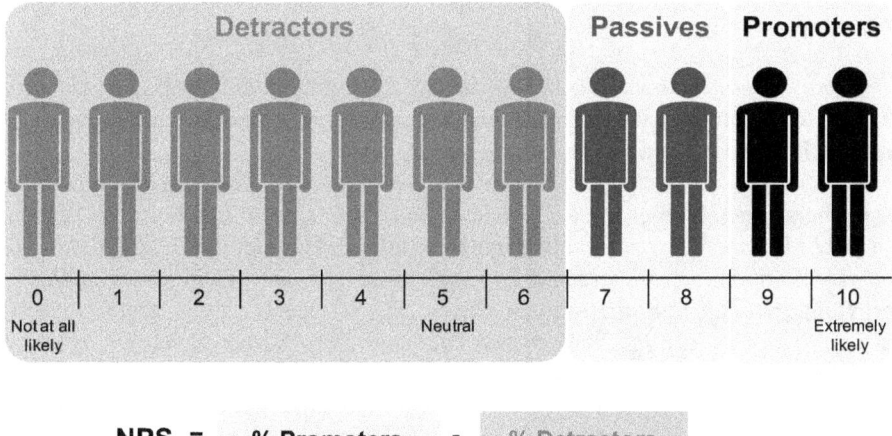

Fig. 6.5 Basic concept of the net promoter score (*Source* Author illustration)

▶ Remember Box The **Net Promoter Score** is a simple and fast instrument for determining confidence—measured on the degree of readiness to recommend to others.

With that, an important factor is the first value determined records the **reference measurement** of the company. In-depth analyses should find out why this particular value was obtained and which measures may be able to improve it. It is essential to, on the one hand, check whether the equation "value of the company's performance for the customer" and the "value of the customer" needs to be improved in the case of an imbalance. On the other hand, a comparison of the competitor environment needs to be made in order to work out possible advantages and disadvantages of the company's position. The NPS concept should become a **basic instrument for the management of customer relationships**.

Think Box
- Is the NPS used in my company?
- If yes, what do we do with the scores obtained, i.e., how do we use them to optimize the abovementioned "score equation"?
- If no, who can be given the job of the initial implementation of the NPS and when can the first results be expected?
- At the same time, how can it be ensured that the scores can be interpreted in the competitive environment?

Trusting Relationship

Fig. 6.6 Establishing a trusting relationship (*Source* Author illustration)

After having elaborated the significance of a learning relationship, we now have to go one step further towards sustainable customer relationships. It is now about **establishing a trusting relationship** (cf. Fig. 6.6)!

It no longer suffices to establish mere relationships with potential customers and customers, whereby they are to be replenished with more information step by step. In fact, it is imperative to **establish** more **confidence** from level to level, in order to be able to submit more targeted and thus more relevant offers in particular by providing suitable information and convincing services (cf. Peppers & Rogers, 2012). With this in mind, the greatest challenge is to further develop one's own customer touch points into **customer trust points** (cf. Fig. 6.7).

The tasks of customer touch point management defined in Chap. 2 must thus be suitably supplemented. The key question is to what extent it is actually possible to establish a **trusting relationship** within the various touch points. Personal encounters such as contact with the employees at the POS but also in a customer service center have the most lasting effect on the relationship between a customer and the company, as experience shows (cf. Keylens, 2013, p. 8).

> ▶ **Remember Box** The customer touch points must be turned into **customer trust points** in order to accommodate for the relevance of trust in the relationships between our company and our customers.

Think Box
- How precisely do we know whether we can really establish customers' confidence at our customer touch points?
- How can we monitor this establishment of confidence and turn it into a key task for all customer touch points?
- What are "confidence indicators" from a customer's point of view such as seals of approval, test results, personal addressing, customized offers, etc.?
- How can it be ensured that all employees comprehend the relevance of this new challenge and implement it in a corresponding manner?
- Where can the corresponding responsibility for "establishing trust" be assigned to?

Another way of quasi-institutionalizing the trust of customers is the **formation of client boards**. For this purpose, client boards are brought to life in innovative companies where two or three times a year, strategic projects, concrete products and services, or the expectations of one's own customers are discussed with selected

Fig. 6.7 From customer touch points to customer trust points (*Source* Author illustration)

Customer **Touch** Points
Customer *Trust* Points

customers. Yet, if in this case only major customers or strategically important customers are involved, one must be aware of the inadequate representativeness of the insights gained for the entirety of the customers.

HCL Technologies, the Indian provider of Information Technology (IT) and consulting company, uses its **Customer Advisory Council** program to gain a better understanding of its customers' business drivers and the major industries in which they operate. It also supplies the company with regular feedback and market insights from customers who help to strengthen and better align its solutions with market requirements. The Council encompasses 110 of *HCL's* Fortune and 500 C-level Customer decision-makers who are contributing significantly to *HCL* annual revenues. Therefore, the council helped to design several new solution frameworks, a better customer satisfaction (25 %), as well as provides a platform, where customers exchange ideas and best practices (cf. HCL, 2014). Many other companies—such as *Dell*, *Nestlé*, *SAP*, *HP*, and *IBM*—have also set up a customer advisory board, in order to create the **requirements for establishing trust** by means of a stronger connection to the end customers.

It is the named examples of customer advisory boards in particular that underline the fact that **confidence as the new currency in marketing and management** has already become more important. In many cases, this may be a reaction to the social revolution previously described: customers are increasingly using the **pressure of the public** to draw the interest of companies. Which means of pressure do the customers apply doing this? Very simple: the **public** for their questions, accusations, requests, and expectations, for such dialogs are taking place to a greater extent "in front of witnesses." The public uproar of customers is becoming normality, in order to literally drive companies in front of them. An important driver behind this trend is the **mobile use of social networks**. These make it much easier to let off steam—from the direct emotional consternation! Due to the little or even lacking reflection of one's own doing, the quality of the language and content of such accusations will suffer! At the same time, the public forces companies to take a stand for questions and issues promptly and competently—in whatever form they are brought up.

> **Remember Box** The social pressure of the public is forcing companies more frequently and more extensively to respond—whether they want to or not!

However, there are still many companies that set up **hurdles for contacting** so that customers establish as little contact with them as possible unless for reasons of purchasing (yet even then, some of the hurdles are quite high!). Try to get hold of a telephone number or an email address for a major airline or a large brand-name company. If you don't go directly to the company info page where such data has to be stated by law, you frequently end up on FAQ pages or are "shoved" into questioning mechanisms, which are supposed to prevent customers from contacting them. Nowadays, customers increasingly use **their own selected methods of contacting**. By doing so, they increasingly define the way they wish to become active! And these are not always the ways contemplated and prepared by the companies. Customers use *Twitter*, *Facebook*, communities, and blogs to air their frustration—and at the same time present it to an interested public eye.

Think Box
- How easy does my company make it for potential customers, customers, and other stakeholders to contact us?
- Are such attempts to contact more likely to be warded off—or are attempts made to gain important insights from them for the entire company?
- How is the responsibility for contacting customers organized in my company?
- Which staff, which budgets, and which management attention are available for it?
- Do we notice critical announcements in real time?
- And how fast can we respond to them?
- Where does the responsibility for these principles of "establishing confidence" lie?

Social media have made it easier than ever before to get **complaints affecting the public** off your chest on these platforms. That is where the eye of every shitstorm can be found, and it is not only large companies that need to be in fear of them. Yet, the larger and more active a company is, the higher the probability that shitstorms will extend into classical media and, by doing so, develop further explosive power. Companies such as *Vodafone*, *WWF (World Wildlife Fund)*, *McDonald's*, *H&M,* or *Nestlé* can write a book about it!

The **dimensions a shitstorm** can take on can be illustrated by the example of *WWF Germany* from 2012. Here, the credibility of the *WWF* in particular was questioned—and for a fundraiser, it is then all about surviving! The following figures show the dynamics that emerged (cf. Scheer, 2012, p. C1):

- 1,700 questions to *WWF* within 2 days
- Within 4 weeks, there were an additional:
  - 4.6 million contacts on the *WWF Facebook* page
  - 839,000 contacts on *Twitter*
  - 263,000 visitors to the website

It is thus imperative to **prepare companies for shitstorms** because they can be triggered for a wide variety of reasons. But how to react when the **shitstorm** is already blustering and, for example, a rumor—"thanks to social search"—is already included at the top of the autosuggestions of the search engine? In some cases, the sensible thing to do may be to just wait until things have calmed down and—to put it bluntly—wait till "another victim is being chased through the village." If we enter into the discussion too emotionally—and by using the same keywords—we will provide our attackers as well as the algorithms of the search engines with additional, unique, and above all current content with the source being identified as our company. A "feast for the search engines" that we should avoid! Above all, it is imperative to keep your feet still if the allegations are true. The most dangerous thing to do is to take action against unpleasant yet true allegations, apparently well-armed, because when the truth is revealed—and most probably will be—your reputation is ruined.

The majority of the users of social media can be boon and bane in equal measure: they can boost and support what the company offers but also pull it into a negative maelstrom if they put the company on the **digital pillory**. It is imperative that companies are prepared for this so that there is no need for a statement such as "We didn't know how to react" made by CEO *H. Schultz* of *Starbucks* regarding a corresponding attack (cf. Karle, 2010, p. 34). And a shitstorm can hit every company—whether large or small!

▶ **Remember Box** Be prepared to be attacked!

If one analyzes verbal attacks on companies, then it becomes clear that they frequently happen in the evening, when the employees responsible for the online presence of the company have frequently finished for the day. Good preparation for **warding off a shitstorm** looks different! Therefore a general **preparedness for social media crises** is needed. The frequency of social media crisis rises; therefore the companies must be prepared to respond rapidly and in a coordinated way. Contingency plans and defined areas of responsibility are necessary to handle crisis successfully.

Think Box
- How well is my company prepared for a shitstorm?
- Have appropriate crisis plans been developed?
- If not, who is responsible for developing them?
- If yes, are they still up-to-date?
- Has a test run been conducted to determine whether the planned decision-making processes and action plans also work?

Whether a shitstorm has the **potential for a PR disaster** or not can be determined by two questions created by *Andreas Schwarz, TU Ilmenau*, (cf. Scheer, 2012, p. C1):

- How does the **legitimacy of the initiators** of a shitstorm look?
- Which **legitimacy of the issue** itself is given?

The **threat potential** can be determined depending on the answers to these two questions. For this reason, companies are well advised to conduct their own **threat analysis** in order to be able to recognize potential threats as early as possible and thus act. Such a **before-fact approach**—known from strategic management—provides the company with many more possibilities of action than an **after-fact approach**, where the "horse has already bolted."

> **Think Box**
> - Have we ever performed an exemplary threat analysis to recognize from which directions a potential shitstorm can come from?
> - Who are the "legitimized critics" who need to be informed about socially responsible actions in advance?
> - What are "legitimate issues" which could be the foundations for a shitstorm, i.e., in which areas is my company vulnerable?
> - Who can initiate such a threat analysis in my company?

What lessons the likes of *German Telekom* learned from the variety of attacks on their own **quality of service** can be taken from the example of *Telekom helps*. In the past, it was extremely difficult to get hold of a person of contact at *Telekom* when having technical problems—who then also felt responsible! With approx. 90 million help calls per year, the complexity of this issue becomes comprehensible. Today, if you post an error message or look for help from *Telekom* via *Twitter*, you usually receive a publicized answer within minutes to then move to the nonpublic domain for further communication. *Telekom* deploys approx. 30 employees to this job in Germany who, however, can only process a small share of the abovementioned volume of calls. However, *Telekom* has managed one thing: with *Telekom helps*, the company has massively gained in "perceived quality of service" in public. Even the German railroad company *Deutsche Bahn* has expanded its service with *Twitter* and now communicates: "The DB Twitter team will answer all service-relevant questions to do with public transport from Mo-Fr 6–22 & Sa-So from 10-22h!" This has likewise improved the image of the *Deutsche Bahn*.

▶ Remember Box **Demonstrating criticism of companies by customers in public forces higher quality of service.** If companies do not accommodate for this, they will be punished for it in public—

until they leave the field, increase the level of service—or the crowd has found its next victim!

These changes also have another consequence. The quality of service a user will be getting from a company in the future will increasingly be oriented towards the **user's social media status**. Any user who feels at home in social media and skillfully uses the likes of *Facebook* or *Twitter* for his inquiries or complaints to the company will experience a **higher degree of service** than customers who fight their way through classic hotlines or bring up their issues by direct mail. And every company is well-advised to ask itself whether it wants to accept this!

At the same time, one target group is receiving additional relevance: the **new opinion leaders on the net**. This target group, which is not sufficiently identified and thus still unattended in many cases, will gain in significance for companies in the future. The focus here is on the following groups:

- Bloggers
- Active and appreciated communicators in online forums and communities
- Users of *Twitter* with a large and particularly relevant community of followers
- Users of *Facebook*, *Google+*, *Pinterest,* etc., who have a high social status and are linked up to many (relevant) people

Companies need to identify these groups and provide them with information in order to take the wind out of the sails of potential critics as promptly as possible. However, such procedure is not always riskless. Under certain circumstances, the provision of certain information may wake up "sleeping dogs." How could these **digital multipliers** be proactively incorporated? When presenting new products, they could be informed in advance or at the same time as announcing to the classical press representatives and be incorporated in the product launch. In perspective, these digital multipliers will receive more significance for the digital natives than the classical opinion leaders, for, due to the change in the behavior of using media, the latter will less frequently get through to the digital natives.

What possibilities are there for identifying digital multipliers? The relevant currency here is the **extent of social networking**. An indicator for this is the *PeerIndex* with the slogan "We value social." In order to determine this value, the *PeerIndex* requires access to *Facebook* or *Twitter* data, the so-called token. The extent to which users are willing to grant this access again depends on the confidence in this partner. Comparable concepts for **determining the "social power"** are offered on *empireavenue.com* and *klout.com*.

> ▶ Remember Box The **dialog with the public** and the dialog with their new opinion leaders, in particular, not only gain in **complexity** but also in **speed**.

> **Think Box**
> - What significance does my company attribute to digital opinion leaders today?
> - Have we already identified the digital opinion leaders that are of significance to us?
> - What contents can be made available to the digital opinion leaders?
> - How is the change in the mood surrounding my company monitored after incorporating digital opinion leaders?
> - Who is responsible for communicating with them?

When it comes to the overall **discussion about confidence**, we must remember that the **Internet forgets nothing**. It is simply impossible to "**un-Google**" yourself. The degree of difficulty involved in doing so can be vividly described as follows (Peppers & Rogers, 2012, p. 101):

> ... you can't take something off the Internet ... that's like trying to take pee out of a swimming pool.

In view of this situation, we should always be aware of the **effects of our actions** on the **confidence potential of our company**.

Quick Wins

-
-
-
-
-
-
-
-
-
-
-
-
-
-
-
-
-
-
-
-
-
-
-
-
-
-
-
-
-
-

Social CRM: The New Rules of the Game in Leading Customers

7

Before dealing with the particulars of social CRM, the **concept of classical CRM** first needs to be specified. As distinguished from the product life cycle, customer relationship management is based on the concept of the **customer relationship life cycle**. This involves the development of the relationship of **individuals** or a **company** or a corresponding **group of individuals or companies** to a **specific company**. Such an analysis helps us to recognize how this relationship (e.g., as measured by the customer value) between a customer and a company develops over time. We can differentiate three core phases (cf. Fig. 7.1):

- **Potential customer management** (motto: "get")
- **Customer retention and customer development management** (motto: "keep and grow")
- **Win-back or churn management** (motto: "win back")

The phase of **potential customer management** involves initiating a relationship to the company. This phase includes the measures for acquiring new customers under the motto "**get.**" The phase of **customer retention and customer development management** describes the development of a customer over time and which subphases he can pass through during that time. Ideally, these include the socialization, growth, and maturity phases. In addition, there may be phases of risk if the expectations are not fulfilled in the second moment of truth, if negative reports about the company involved or products emerge, or if competitors submit interesting offers. In this phase of customer retention and customer development, companies can apply various measures to retain the customers and exploit higher sales/profit potentials. The motto here is "**keep and grow.**" The transition from customer retention to **win-back management** is affected by the degeneration phase, where the intensity of the relationship decreases and the company is in danger of losing the customer. The motto here is thus "**win back,**" however, only if attractive customer values can be expected in the future.

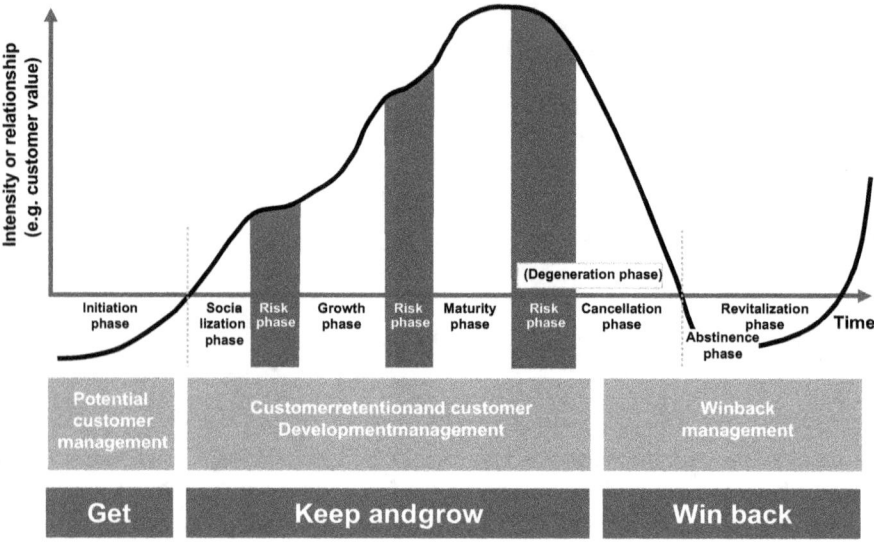

Fig. 7.1 Customer relationship life cycle (*Source* Author illustration, data source: Stauss, 2000, p. 16)

▶ Remember Box Each of the **phases of the customer relationship life cycle** described is accompanied by **specific demands on us as the accommodating company**—for every person and every company. The individualization of customer care must coincide with the expected customer value. For this reason, we should—at least for the most important customers—know in exactly what phase they are at the moment and which future value contributions are to be expected.

Think Box
- What significance does my company give to customer management today—in terms of the concept of the customer relationship life cycle?
- What information is still missing, if any, to apply this concept?
- Which result potentials can be opened up for my company by applying this concept?
- Which possibilities do we have to consistently aim the support measures at the various phases of the customer relationship life cycle?
- Where can the responsibility for the corresponding procedure be placed?

Before **potential customer management** as per Fig. 7.1 can take place, the focus must first be placed on the **acquisition-oriented segmentation** in the scope of the CRM. Based on the acquisition targets and/or knowledge gained by determining customer value (cf. Chap. 5), we define which target segments we would like to approach ("**definition of the targets**"). This determines which target group or target groups marketing and sales should be aimed at. This determination of the **focuses on acquisition** is not only relevant for the **design of the sales and marketing concept** but also for the selection of the **sales and marketing instruments** to be used for the acquisition. Only if we know precisely who we want to reach can the suitable communication channels and instruments be selected and appropriate offers submitted!

Besides the acquisition-oriented segmentation criteria and concepts primarily used for the initial target group definition, a **transaction-oriented segmentation** must be conducted for a company's potential customers and customers already acquired (cf. Kreutzer, 2013 for more details). This draws on the information already acquired in the course of transactions between potential customers and customers on the one hand and the company on the other. The transaction-oriented segmentation thus facilitates a much greater depth and focus on the description and elaboration of the segment than the acquisition-oriented segmentation—if the company has endeavored to acquire corresponding information. In the case of acquisition-oriented segmentation, in contrast, only a little basic data can frequently be relied on.

When looking after people and/or companies along the customer relationship life cycle, we should remember that, on the one hand, the **interests** and the **information needs** of people and companies are pronounced to varying extents in these various phases. On the other hand, we as the providing company follow **various goals** in the individual phases. In the course of **potential customer management** it is primarily about approaching target and desired customers for the company in different ways in the scope of an initiation phase, in order to get them enthralled by our own offer. In doing so, it must be taken into consideration that potential customers exhibit different information needs than standing customers. The transaction-oriented segmentation can draw on information already gathered about the potential customers. In doing so, the knowledge acquired is drawn on for the organization of further customer care and for the optimization of acquisition measures themselves—under constant consideration of the customer value as key control factor (cf. Chap. 5).

> ▶ **Remember Box** It must be ensured that the **target persons** are **taken care of in a differentiated way** depending on their position reached in the customer relationship life cycle to our company.

A **potential customer,** for example, first looks for **sources of supply** for exclusive offers such as watches by *Lange & Söhne* or vehicles by *Bentley*. Or he would like to receive **information about modalities of payment by installments** at *JCPenney* or about the **financing options** for the purchase of a flat screen TV at *Best Buy*. Apart from that, in the acquisition phase, it needs to be found out why a

Lange & Söhne watch is preferred over a product by *Maurice Lacroix* or the range of clothing by *Nordstrom* apparently outclasses that by *Zara*. At the same time, it is necessary to **dissipate** the existing **initial purchase resistance and uncertainties**. This can be successful by means of discount coupons or by general discounts on first purchases. Yet, competent advice and a customized and fast provision of information contribute also towards this. Now, the **relevance of a consequent customer touch point management** again becomes visible (cf. Chap. 2).

The **customer retention and customer development management** phase comprises several levels which, in turn, make various demands on corporate marketing and, in particular, on the communication tools to be applied (cf. Fig. 7.1). In the **socialization phase** the customers are first to be made familiar with their new service partner. On the consumer market, these are, for example, the car showroom, the chosen clothing company, the website of an online wine mail-order company, or a fundraising enterprise such as *UNICEF* providing customer care. In the field of B2B this phase proves to be significantly more challenging in the case of investment goods, for example, if the user has to be familiarized with complex ERP software from *SAP* or a new printing machine from *Heidelberger Druck* as well as with the companies and employees (sales, service, training) behind them. Under ideal circumstances, this process follows a **growth phase**, in which sales then increase when the customers have learned to trust their new provider or service partner and avail themselves of additional services. The **maturity phase** can be reached within a few days, weeks, and months of years, depending on the offer.

In addition, the company should consistently initiate the **triad of customer care** if the appropriate products or services are available. In the course of **selling more**, one should try to create customer loyalty so that the customer sticks to the product or the provider as in the company providing customer care and regularly buys "more of the same." This is an approach by customer retention systems with which repurchases are rewarded (e.g., by all the frequent flyer programs). On the B2B market, a specific way of intensifying the cooperation between manufacturers and specialized trades or tradesmen can be established to achieve this objective. By means of **cross-selling** we motivate the customers to also make purchases in other areas of offers of the company. In the case of *AmericanExpress*, this would be the likes of making use of other financing services such as insurance or overdrafts. It is thus aspired to get a higher turnover from one customer address—or, to be more precise, to achieve a higher contribution margin—and thus a higher customer value. The measures of **up-selling** are thus aimed at motivating the customer to purchase higher quality products of the same company. In the case of *AmericanExpress* this means the likes of a customer having the green credit card being regularly offered the gold card which triggers a significantly higher contribution margin for the company.

This ideal type of development can be consistently interrupted by the previously mentioned **risk phases**. This may be, for example, aggressive price offers, new sales policies, or innovative ranges of services from competitors, service breakdowns, or a failure in social media. The more stable the customer relationship is, the more difficult it is for new providers to penetrate existing relationships. Here, a certain parallelism becomes apparent between the interpersonal relationship networks!

The last phase in the scope of the customer relationship life cycle is **win-back management**. Here, we should first try to identify those possibly going to cancel but ahead of them doing so, in order to prevent customers from leaving. A particular challenge for companies comprises identifying such customer losses if there is no contractual relationship to be canceled. This is given, for example, in the most diverse types of the trade. Here, customers do not cancel but become "inactive." And only those companies that have information about ongoing sales, for example, based on a customer retention system, can find out such inactive customer. Yet even them—according to our experience—frequently do not do this!

If a customer has become inactive, then it's time for **churn management**. This describes all activities that attempt to "turn the customer around" again so that he overcomes his inactivity or withdraws his notice. If a customer has turned their back on us, then the question must first be posed whether he should be won back, for not every customer deserves us fighting for him to stay! In the case of particularly important or valuable **inactive** or **terminating** customers, it should be systematically attempted to argue them out of their decision and to persuade them to remain with the company. Those canceling their mobile phone contract or newspaper/magazine subscription frequently "enjoy" intensive win-back management. Particularly affordable offers, advantageous conditions of contract, give-aways, and others are the incentives to motivate a customer to stay. Other possibilities are coupons with savings, invitations to VIP events, or small presents to win back customers. However, a requirement for the target-oriented use of such measures is up-to-date and meaningful information in order to be able to undertake value-oriented customer care that is actually received with great interest.

> **Think Box**
> - What possibilities can my company make use of in order to achieve differentiated customer care of potential customers and customers across the various phases of the customer relationship life cycle?
> - How can we measure which additional contributions to the results are to be generated in order to set them off against these additional customer care costs?
> - How consistently is the triad of customer care put into practice in my company?
> - Is the view over the brim of the teacup given for the intracompany exploitation of customer potentials or does a silo mentality hinder such an approach?
> - Can we recognize when customers become "inactive" and is this communicated on a regular basis?
> - Have we set up a consistent win-back management system?
> - Where is the responsibility for the management of customer relationships best placed? Should it be in marketing or sales or better still in an overlapping entity?

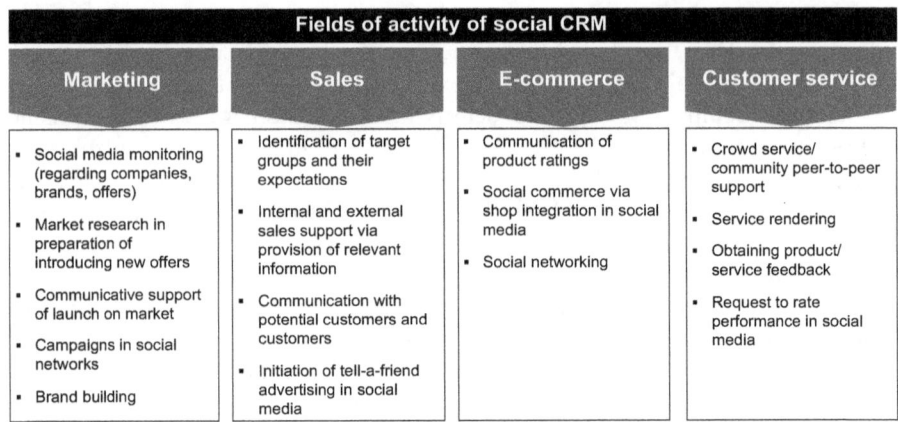

Fig. 7.2 Fields of activity of social CRM (*Source* Author illustration)

This information emphasizes that CRM has always been aimed at the organization of social relationships between companies and their customers. Today, however, there are good reasons why social CRM is being increasingly spoken about and should be. As already mentioned in Chap. 2, with **social CRM**—accompanying the already established concepts—social media services and social media technologies are used to further enhance the interaction between customers and companies. Figure 7.2 illustrates the bandwidth of the activities belonging to social CRM. In part, fields of activity which have also been mentioned elsewhere have been allocated.

For us companies, the **establishment of intensive customer relationships** represents an important target of CRM. With that, the question must be posed as to which (social) media are particularly suitable for this. In Fig. 7.3, three **media categories** are differentiated. First of all we have the category **broadcasting**. This is where frequently relevant one-to-one or one-to-many communication still takes place in the external media as well as in TV, radio, and print. The **intensity of the relationship** achieved by this is thus low. As well as that, it is becoming increasingly more difficult to reach target groups via these classical media, because users often channel-hop or spend little time on printed advertisements. Other media aim at a higher extent of **engagement**. Platforms and media offers such as *Pinterest*, blogs (like *Tumblr*), communities, and *YouTube* or *Vimeo* demand and facilitate a much higher degree of engagement (as in targeted searching) or contribution as such (by writing and/or uploading own content). Social networks such as *Facebook*, *Google+*, *LinkedIn*, and *Twitter* are present in both one-to-one, one-to-many, and also many-to-many communication. Even mailings and emails are assuming a hybrid position here, because they can reveal a one-to-many and a one-to-one extent. The most personal form of interaction—with the greatest achievable intensity of relationship—still must be the telephone call and in particular the conversation at the POS.

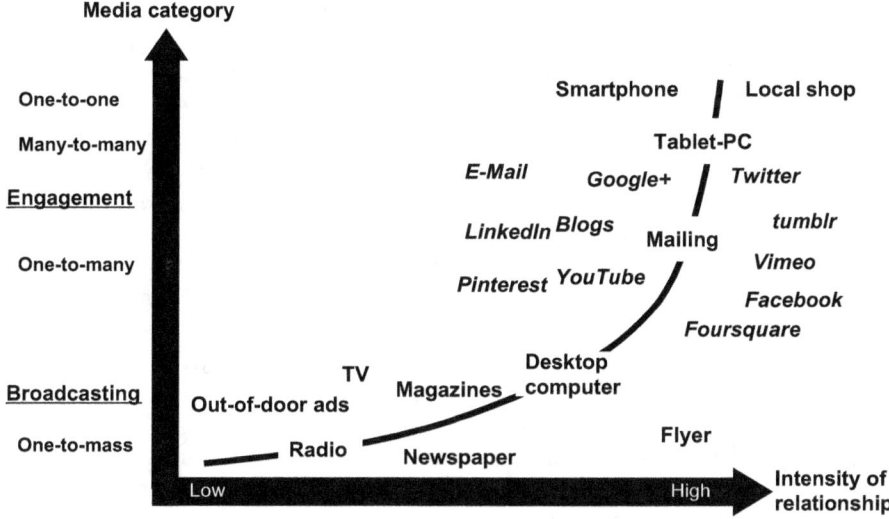

Fig. 7.3 Intensity of the relationship according to the media category (*Source* Author illustration)

> ▶ **Remember Box** Anybody who wishes to establish a great intensity in the relationship with his customers frequently cannot get past social media.

> **Think Box**
> - Have we ever given thought to what intensity of the relationship to our customers we wish to aspire—differentiated according to customer value where necessary?
> - Have we ever assessed our previous instruments of communication for the possibility of establishing an intensive relationship?
> - To what extent can we increase the profitability of our own actions by which customer loyalty increases or by means of an increase in repurchase quotas or an increase in the average sales slips?
> - Who is responsible for these issues?

The **requirement for a high intensity in a relationship** is in many cases the **knowledge of the customers' concerns**. For this reason it was and is a key task of classical CRM to derive indications of the customer's preferences from various data and to do so as close as possible to the scheduled time of purchase. Complex systems have been developed to achieve this objective. However, all companies were adversely affected by the **cost-intensive data maintenance** needed for this, because the **data**—in particular with regard to purchasing preferences—only has a **short half-life value**. A particularly interesting solution to this is being discussed in the context of social CRM: the **access to the information available in social networks**. The information stored there offers a gigantic potential to update the information in one's own database and, in particular, to extensively add to it.

Facebook, for example, today with more than 1.3 billion participants, represents the largest **profile and preference database in the world**. It is updated by millions of users day after day and designed with "love for detail." This is why *Facebook* is also the best maintained database. The problem of many databases until now—the short half-life value of data—has thus been overcome by this. The contents of posts, changes in relationships, and declared Likes are altered in real time and can be evaluated in real time. *Facebook* offers another special feature: as companies were only able to communicate with IP addresses in those cases where no log-in was necessary or where no email address existed, **open graph** now facilitates the access to real user profiles (incl. interaction data). In the case of *Facebook*, this open graph covers the **interest graph** with information about preferences as well as the **social graph** with the social relationships structure. The social graph makes the **relationships of the customers among each other** analyzable and provides interesting potential for viral processes.

Access to this data is allowed if customers permit the companies to access their open graph. To do this, the customer must grant permission (also called **token**) to access data. Only then can *Facebook* allow analytical access to this data and enable the company to feed the user with relevant information, instead of "spamming" it! This token can be gained in three ways:

- In **own web applications**, where a social log-in is made use of
- In the form of **apps**, promising special services
- On the *Facebook* **page** itself

The *Facebook* API enables the "open sesame"—with access to the user, his full profile (incl. photos, events, pages), including access to the user's network. By this, *Facebook* provides a previously unknown **depth of information**. This depth of information has not yet been achieved by any CRM database—however well maintained!

What exactly do we understand by a *Facebook* access token? An **access token** represents a series of **user rights** that can either be exported via *Facebook* or imported back to *Facebook* (cf. Wohlfarth-Bottermann, 2013a). The exporting of a token requires the express permission of a user via an application ("creating permission as opt-in"). Each application user has its own access token. A token comprises a long text string that cannot be deciphered by man (cf. Fig. 7.4).

The **exporting of a token** typically only allows a restricted view of the data of a *Facebook* user. In order for an application to be allowed to collect data, the user must allow access to his data. This includes individual areas that the application may access. In this way, an application can export profile information, level of education, birthday, professional profile, relationship status, the *Facebook* wall, and/or information about friends. Figure 7.5 shows how the **permission to access data can be**. Technically speaking, every application has the possibility of

Fig. 7.4 Token format in an excel sheet (*Source* Author illustration)

Required Permissions	Recommended Permissions	Nice to Have Permissions
Access user´s basic information • Name • Gender	**Access user´s profile information** • Hometown • Interests and activities • Education and work history • Political views, orientation, quotes • Events, RSVP´s, and groups	**Access user´s profile information** • Religious views • Friends • Friendlists
Access user´s profile information • Birthday • Languages • Likes • Music, TV, movies, books, games • Current city	**Check-ins** Application may read user check-ins and friend´s check-ins	**Access user´s family and relationships** • Significant other • Family member relationships
Access user´s family and relationships • Relationship Status	**Access posts in the user´s news feed** • Links, tags • Notes, posts • status	**Access user´s friend´s information** • Birthday • Likes, activities and interests • Music, TV, movies, books, games • Hometown, current city • Education and work history • Religious and political views, quotes • Events and groups
Access user data anytime App may access user data even when not using the application		

Fig. 7.5 Facebook data an application can access (*Source* Author illustration)

retrieving the entire token of a user. However, this is not always constructive as the depth of the token retrieval can also increase the disconnection rate when using an app. The *Facebook* guidelines stipulate that an application may only retrieve the information necessary to implement the app (cf. Facebook, 2012b).

We can illustrate the implications on the basis of the *JCPenney* catalog. The main catalog has hundreds and hundreds of pages and thousands of articles. But how many of them are really interesting for the respective recipient? A single person skips the children's stuff, the occupant of a multistory building skips the garden section, and the "exercise-hater" all corresponding products. Therefore, each individual user is frequently only interested in a small part of the product range. So why go to all the effort when the **precise and up-to-date customer preferences** are available by accessing the *Facebook* **interest graph**? An increasingly finer segmentation with more precisely offers and "min catalogs" matching the target persons—one-to-one where possible—are becoming possible and the costs of which need to be reviewed.

The **silver bullet** for the connection between companies and users is **permission-based**. One needs to make the customer realize why he should grant access to his data and of what benefit it is to him! What could they be? On the one hand, the user would like to live the "social experience" of *Facebook* with other applications, too. The **social log-in** via *Facebook* would be appropriate for this. When using it, the data maintained—frequently with much love for detail—on *Facebook* do not need to be entered again in other applications. And an update on *Facebook* automatically leads to other applications being informed about the change. On the other hand, apps that access the depth of the *Facebook* data can achieve **a hit accuracy of information and offers** that wasn't even perceivable in the past. *Hallmark*, for example, offers a **social calendar** as a service via a *Facebook* app with the birthdays of friends. On the basis of this, suitable gift ideas can be presented at the same time (cf. Eckerson, 2012).

Here the **relevance of the equation of less privacy in favor of more one-to-one offers** becomes apparent. And the basis required for this is **mutual trust**. This is why we must always give the user the feeling that he himself is controlling the processes based on his data. We are still at an early stage. But sooner or later, the journey will take us there—to the union of both worlds!

The *Facebook* token has another important **multiplier effect**. If every *Facebook* user has an average of more than 300 friends, 300 further sets of data can be accessed via the social graph in different ways, depending on the scope of the permission. Via 1,000 tokens, data from an average of 300,000 people can thus become analyzable. This again reveals what great significance the **trust** in the information and service partners will adopt. Companies trained in analyzing can derive the diversity of this information with further details—for example, about the "estimated income"—based on the profile data. By doing so, the legal restrictions in each country have to be considered.

> **Food for Thought** Big Brother is watching you—and you know it. And you may like it!

Figure 7.6 illustrates the way traditional CRM differs from social CRM with regard to essential aspects. First of all, it can be noted that both the **setup costs** and the **maintenance costs** of traditional CRM are a significant cost factor. In the case of large companies, totals amounting to several million US $ are frequently reached. Even with the **bandwidth and depth of the base data,** there are significant differences. As in the case of social CRM, the users maintain the data themselves and the question of **half-life time** does not apply. For if users of *Facebook* do not look after their data well enough, then they will have to do without relevant news and offers—or they are requested by "friends" to do an update or correct "wrong data." One thing is becoming clear: whereas traditional CRM can work mainly detached from the **trust of potential customers and customers,**

	Classical CRM	Social CRM
Setup costs	Usually > $ 100,000	Free
Costs for maintenance	Usually > $ 50,000	Data maintenance by user and network operator (e.g. *Facebook*)
Width and depth of data	Rather narrow and flat (depending on meticulousness of company)	Very deep (depending on type of token)
Half life of data	Ø 24-36 months	Continuous maintenance by users
Required extent of trust	Low to average	High (correlates with type of token)

Fig. 7.6 Comparison between classical CRM and social CRM (*Source* Author illustration)

social CRM is then again based on the "currency of confidence." At the same time it holds true: due to the increasing prevalence of *Facebook,* an ever **more powerful communication channel** is being established which, in comparison to traditional media, is gaining continuously in reach.

Think Box
- What possibilities are there for my company to obtain access to the interest and social graph of *Facebook* via my customers?
- How can this improve the structure of my fans and customers?
- Which of the information available here enables my company to submit even more relevant offers?
- How must the acquisition of the *Facebook* token be designed?
- In which field of responsibilities does the matter belong?

Another interesting development of **social CRM** comprises the achievement of an intensive **integration of the customers in intracompany processes**. These can refer to fundamental product innovations and the development and selection of advertising slogans to the (co-) creation of entire communication campaigns, as already presented in Chap. 4. The use of so-called join-in Webs can integrate a large number of users that could hardly be activated in the past more comprehensively in company processes. This is why there are a variety of new trends and possibilities having the prefix "crowd." This is the case, for example, with the so-called crowdsourcing—true to the motto: "It's the mass that counts!" In this connection, people—not necessarily customers of a company—are motivated to put forth ideas and proposals in company processes. If this integration takes place in production processes, it is also called **peer production**. While *Marc O'Polo* used the wisdom of the crowds of his customers to expand his own shop, *Henkel Beauty Care* actively integrates their customers in the product development; with that, direct contact is of the utmost priority. In this way, a variety of instruments are used to learn interesting facts about the application of beauty care products from the horse's mouth. In addition, the company division is increasingly engaged in providing customer care services at the customers and focuses intensively on Web 2.0 applications.

By using the most important social network—*Facebook*—as an example, the following shall illustrate which exciting **areas of use of the social networks** can be found here. At the same time, the limitations of the use will be made clear. First of all, we must be aware of the fact that *Facebook* represents the private living room of many of our customers, because *Facebook* focuses on the private dialog. Can we, may we, or should we "intrude" here? Or isn't it rather a matter of becoming esteemed and "beloved" partners of conversations and dialogs for the users?

At this point, a frequent misunderstanding regarding the engagement in *Facebook* needs to be done away with. The **fan page** on *Facebook* does not represent the anchor of communication from a user's point of view as may be the case with a website. For many fans, it is more the case that after the first visit to the fan page and "becoming a fan," many *Facebook* users never visit this fan page again, even though they have become a fan.

> **Remember Box A Like is not enough!** The aim of engaging in *Facebook* doesn't stop at collecting fans—on the contrary. Anybody who has gained many *Facebook* fans is virtually obliged to hold something for them—for engaging in social media and *Facebook* are like an invitation to a party. It then has to take place!

The most frequently determined failure is illustrated in Fig. 7.7. **Becoming a fan** represents only a **minimum commitment** of the users, which is not very resilient. Anybody who solely trusts in a large fan base leading to a **high reach** and to an **increase in the ROI** will end up disappointed. The reason lies in the dynamics of social networks. In order to get communication going and keep it going, interesting content is necessary. Once again it must be emphasized: the fan page of *Facebook* is a **no pull medium**. The establishment of relationships and thus the traffic between companies and users are achieved via the **posts** we as a company provide!

We must therefore make the **four jumps to success in social media** clear to ourselves (cf. Fig. 7.8). These four jumps to success also begin with the **winning of fans** as the **1st step**. Yet, ideally, one should not only be tempted to win fans; it is more important and essential to gain the previously mentioned token, in order to be able to access the depth and breadth of the *Facebook* data. A merely feigned implicitness should not be lost sight of when trying to win fans. These are the answers to really simple **questions from the users' point of view**:

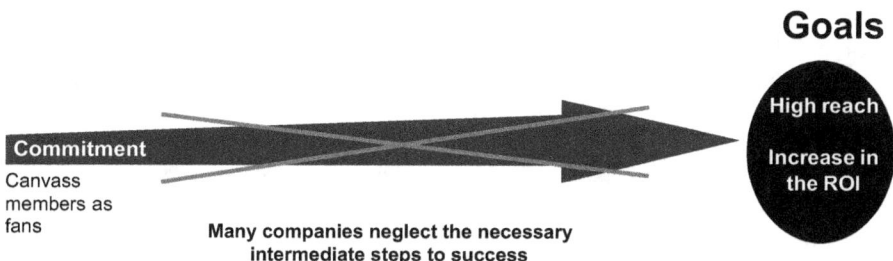

Fig. 7.7 How a Facebook engagement can come to nothing (*Source* Author illustration)

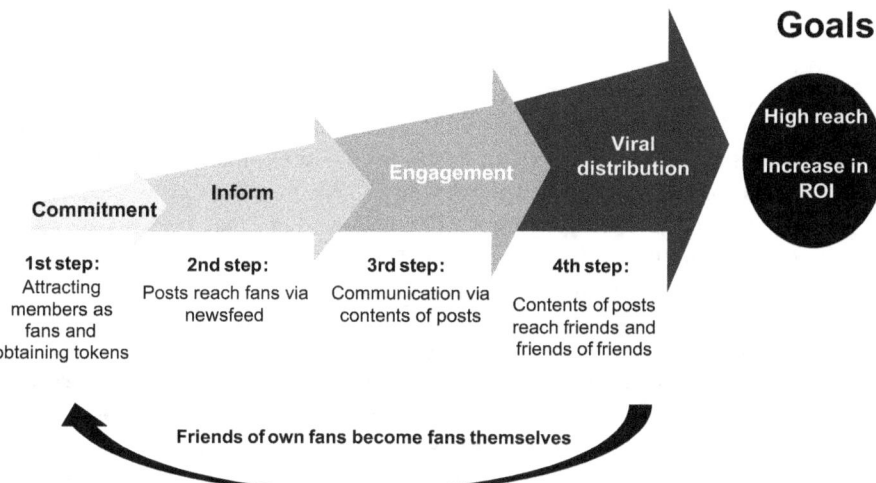

Fig. 7.8 Four jumps to success in social media (*Source* Author illustration)

- Why should I become a fan or follower?
- What's my reward?
- Why should I spend my time on it?
- What's in it for me?

Amazing how few companies have been able to provide a comprehensive answer to these questions.

After the customer has committed himself to a certain extent by becoming a fan, the **2nd step** is to maintain the trust potential by relevant **posts** to the fans. With this, it becomes obvious: the relationship between the company and the fan is shifting from the fan page to the—hopefully exciting, interesting, new, and thus relevant—**company news**, which appears in the fans' newsfeeds. With that, it must be remembered that these posts—as many have wrongly assumed—are not sent to

100 % of one's fans. The number is less than 6 % and still decreasing (cf. Löhr, 2014, p. 22). Insiders even assume that frequently, only 2 % of one's own fan base can see the **posts**. What solution does *Facebook* offer here? **Paid posts**. In this case, the companies have to pay for their posts actually reaching the majority of their fans—in consistency with the leitmotif of *Facebook*: "Free and always will be!" Why can *Facebook* allow itself to be paid for the delivery of messages to one's own fans? Because the fan community set up by every company is legally owned by *Facebook*!

The central leitmotif when creating posts is **relevance**. Yet how can such relevance be generated? This requires the **three Cs**:

- **Contact**
- **Content**
- **Context**

Via the newsfeed, the desired **contact** to the fan is established. The **contents** presented in it are to ideally be integrated in such **context** that the announcements are important for the user just at that moment. This is how the desired relevance is developed. With that, all companies—as demonstrated in Chap. 6—should systematically canvass the users' **trust**. This results in the development of a new "currency" that can be abbreviated to CCCT (cf. Fig. 7.9).

The aspired relevance is also the requirement for the desired **engagement**, so that in the **3rd step**, our fans discuss the communicated contents in their respective network (cf. Fig. 7.8). It may be possible to turn *Facebook* into a **social recommendation engine**. With that, the requirement for a **viral distribution** of our contents has been achieved in the **4th step**. For, if the friends of our fans are enthusiastic and in turn share the contents, then we will experience **second-order virality**, because friends of friends hear of our activities. They can also become fans. This turns *Facebook* into a **social recommendation engine**. The engagement can also fortify our original fans in having engaged in the "right contents" (ours!). It is not until these manifold intermediate steps do we reach a **high reach** and can—due to the relevance of our messages—work towards a **positive ROI of our activities**.

In order to achieve a positive ROI of the *Facebook* measures, we must, however, bear one thing in mind: it cannot be the permanent task of the companies to permanently "entertain" the customers for them to participate. It is much more important for us companies to give **impetus for self-engagement of the fans**. Therefore a critical number of *Facebook* fans is required.

Napkin Laps provides an interesting study of **US use behavior** on *Facebook* with regard to the extent of **fan engagement**. To this end, an analysis was conducted based on 52 brand name pages on *Facebook* for 8 weeks as to what

Fig. 7.9 Creating a new currency in marketing (*Source* Author illustration)

CCCT –

Contact Content Context Trust

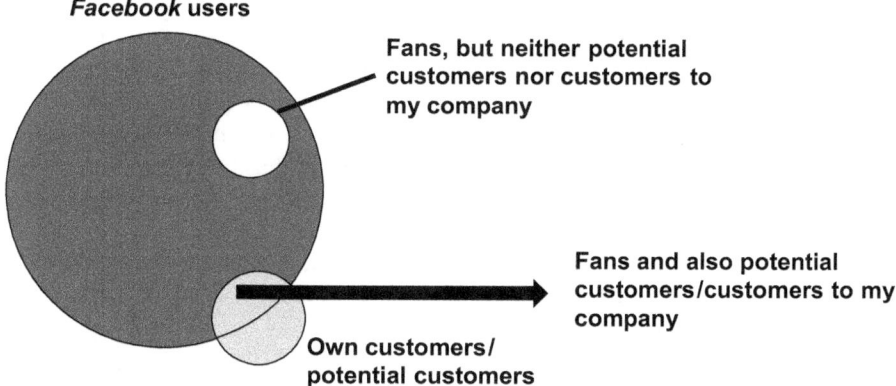

Fig. 7.10 Which Facebook fans are also potential customers/customers to the own company? (*Source* Author illustration)

extent the users got involved. The fan pages taken into consideration in the study included between 200,000 and one million fans so that, in total, the behavior of more than 30 million fans was included in the study. The main result of the study is **6 % of the fans engage on average**. In addition, a very interesting result was revealed. As many brand names are obviously only interested in winning Likes, they lose sight of looking after the fans they have already acquired (cf. Lafferty, 2012). And this should be avoided.

With the **engagement** in *Facebook*, an additional and important if only rarely posed question unfolds: Are the *Facebook* fans "gathered" also potential customers or customers of the own company? Figure 7.10 illustrates this question. The **gross reach in social media** disclosed in many company reports and statistics merely states the number of fans or followers irrespective of whether they are at the same time potential customers or customers. As not every fan is also a potential customer or customer, one should try to determine Net reach **in social media**, which represents the overlap between *Facebook* fans and the own potential customers/ customers. Consequently, not every fan who "outs" himself as a fan is of the same significance for the company under the aspect of sales—in the short or medium term! Nor is every one of them an opinion leader or influencer needed to be taken great care of. At the same time, however, one shouldn't punish those with a lower customer value with disregard. Appreciative behavior is thus most advisable here, too!

At the same time, Fig. 7.10 also clearly defines the actual task. It is imperative to **win** as many **potential customers** and **customers** as well as relevant **digital opinion leaders** and other **influencers as fans** as possible, in order to predominantly involve precisely those in the communication regarding the company, its brands and offers! For what use are posts to people who cannot afford my products or don't want to and otherwise do not (positively) contribute towards establishing the image? This group of people represents the interested public for the actual buyers of my products at the most!

The question about the **overlap between the own potential customers/ customers and the fans** must be responded to in particular, if the CEOs or CMOs mainly look at the number of *Facebook* fans without considering the extent to which a **relationship exists between the fan community and possible buying activities.** For one thing is sure: The more aggressively competitions are used to build up their fan base, the more attractive, generic, and thus further away from the brands the promised incentives are, the larger the number of the correspondingly addressed opportunists will be, who are not really interested in the brand being promoted or maybe cannot afford it! Apart from that, it must always be checked whether fan-winning campaigns are compatible with the *Facebook* **promotion guidelines**—which change from time to time.

Furthermore—in particular to the CEO and CMO—the following message should be sent:

How many fans a company has is not important!

Who we have won as fans is more important—and whether they also **engage** themselves on behalf of the company. For this reason, it must also be noted: not only is it not worth it, but it even damages the consistent presence of a company to buy fans. Companies such as *usocial.net* offer the likes of *Facebook* fans and *Twitter* followers for sale!

Purchased fans distort the profile of the own users in the same way that fans, who bear no relation to one's own range of products, are attracted by competitions. All of these measures do indeed increase the number of fans but not the **extent of the relevant fan community**. What brand would like to count competition junkies and opportunists among their fans? For even for *Facebook* fans the following again applies: **quality before quantity!**

Only with a **relevant fan community** is it constructive to differentiate between two different types of reach. **First-degree reach** revolves around one's own fans. An additional important indicator is, however, the **second-degree reach**, with which the number of the friends of the fans is meant (cf. Fig. 7.11). For even in these networks, communication can take place if, for example, a fan of *Audi* likes a post, because this information turns up in the friends' news stream. Every company is well advised to target the collection of second-degree reach.

> **Think Box**
> - What significance does my company place on the mere collecting of *Facebook* fans?
> - How well is the "four jump approach" implemented by us for success in social media?
> - How consistently do we make use of posts or is our main focus on our *Facebook* fan page?
> - Have we ever tried to determine our gross and Net reach in social media?
> - How high are the first- and second-degree reaches of our fan community?
> - Where is the responsibility for the handling of these issues?

7 Social CRM: The New Rules of the Game in Leading Customers

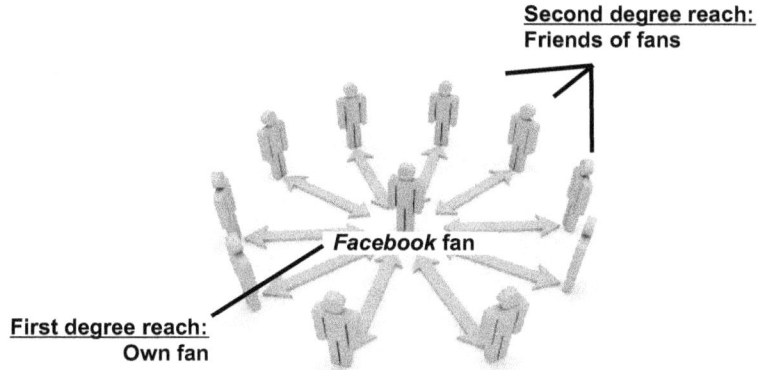

Fig. 7.11 Determining reach in social media (*Source* Author illustration)

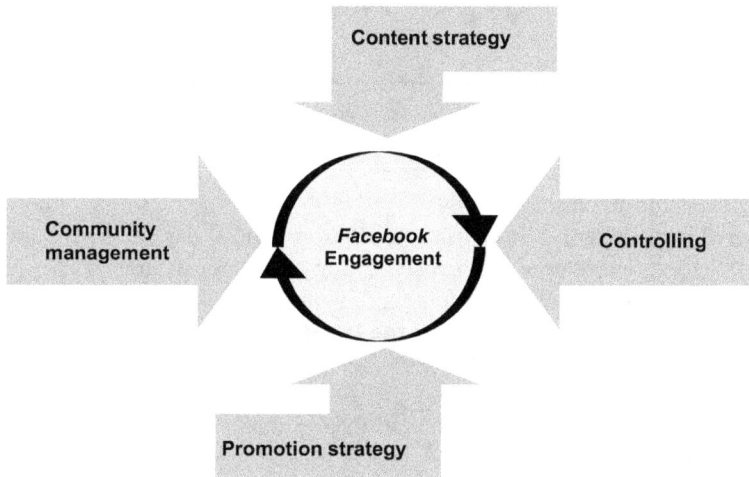

Fig. 7.12 Fields of action of a Facebook engagement (*Source* Author illustration)

The **key issue of the use of** *Facebook* that every company should answer for itself is: How can we win our own potential customers and customers as fans and contribute towards the sustainable value creation for our own company? Figure 7.12 illustrates the four relevant **fields of action**.

First of all, a consequent **community management system** is needed. The first task of it is to define the goals for the *Facebook* engagement (cf. Chap. 9). In addition, what interests and expectations there are in our core target group with regard to the contents provided must be determined at the beginning and then continuously. If the users permit an analysis of the interest graph via the *Facebook* token, then a more extensive consideration of the (future) interests of the users can be achieved than was the case with traditional CRM concepts.

Based on this information, the **content strategy**, which revolves around the company, its brands, and offers and, ideally, exciting stories about these, needs to be elaborated (keywords: **storytelling** and **narrative marketing**). Even an **advisory approach** is becoming increasingly important in order to help the customers solve everyday problems. For this reason, it is imperative to recognize the customer's "problems" and "needs" from a recipient perspective, in order to provide appropriate contents. This content strategy can begin in classical media and be extended into the online area or vice versa. Yet, the content strategy can also be fed from the contents of social networks themselves. In this way, the cosmetics provider *Kiehl's* seizes the beloved formulations of its customers and incorporates it in its advertising communication.

In principle, it has crystallized that the following **areas of a content strategy** within the social networks are regarded by fans of the company and brands as particularly attractive and motivate a visit to the social site of companies. The following results presented reveal the **expectations from a consumer point of view**—based on a global study by IBM (2011b, p. 9):

- **Discounts** (incl. special savings, limited offers that can be aimed at different target groups)
- **Purchases** (e.g., by means of links to online shops or referrals to stationary locations)
- **Feedback and product ranking**
- **General and exclusive information** about products, the company, and/or the industry (e.g., via referrals to websites, blog entries, or forums)
- **Information about new products** (e.g., with launches)
- **Expressions of opinion regarding the latest products/services** (e.g., as dialog platform for the exchange with other users)
- **Customer service**
- **Participation in events** (incl. invitation to product or company presentations such as fashion shows)
- **Feeling part of** (being part of a larger community)
- **Expressing ideas for new products/services** (such as idea competitions)
- **Belonging to a community**

Figure 7.13 shows how far apart the **expectations of the consumers** and the **estimations of the companies** lie. Even if this study is only based on a small random sample, trends can be defined at the least. Whereas **discounts** and **purchases** are right up front of the consumer ranking, these take the last places when it comes to the decision-makers. Companies should in fact avoid to only "presume" what their customers would like to get!

In view of these results, when designing these activities we should keep in mind that *Facebook* in the first instance is **not an additional sales channel** but a **communication channel**. This becomes vivid with the following quote: "Instead of providing relevant information in context, answering questions directly and being helpful, many marketers use spam slingers on *Facebook, Twitter* etc. and

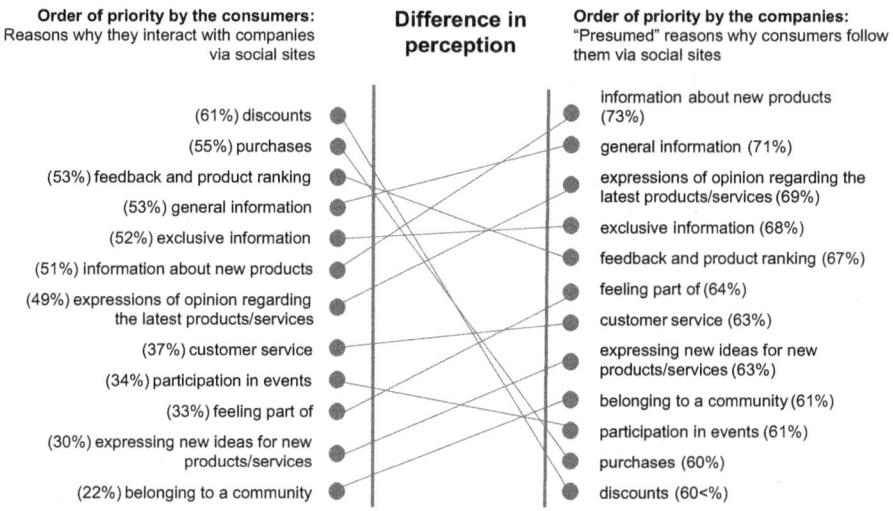

Fig. 7.13 Discrepancies between companies and consumers regarding the reasons why consumers communicate with companies via social sites ($n = 1{,}056$ consumers, $n = 350$ - decision-makers, global study) (*Source* Author illustration, data source: IBM, 2011b, p. 9)

wonder why their measures don't pay off" (Steimel, 2012). Companies should avoid turning *Facebook* into such a **spam-slinger**, if they want to build up the attentiveness and confidence of their target persons and maintain it in the long run.

Based on the mission statement "content is king" and "relationship is queen," another field of action of the *Facebook* engagement in the course of a **promotion strategy** is to achieve the activation of users so that they do actually make use of the abovementioned offers (cf. Fig. 7.12). To do this, besides the *Facebook* posts themselves, various online tools (such as banners, blog entries, referrals in communities) and offline tools (such as classic advertising, posters, or TV spots with corresponding referrals) can be employed. As well as that, *Facebook* offers the possibility of **placing** *Facebook* **ads**, whereby the circle of recipients can be defined very precisely on the basis of *Facebook* data. Another important sales-supporting function of *Facebook* can be seen in the prompt to participate in **cross media traffic**. This means, for example, in the scope of your *Facebook* presence, providing links to your own *YouTube* channel, your own online shop, or off-line presence—for example, as a stationary retailer.

An important contribution to promoting *Facebook* is also made by the large number of **social plugins**, by means of which one's own *Facebook* presence on the Net can be extended. These social plugins include the Like button on *Facebook* as well as the recommendations and also the *Facepile plugin*, which shows the *Facebook* profile pictures of the users who have connected to a website. These applications are called social plugins because they facilitate a **social experience** for the users without having to leave the website.

Finally, all measures taken are to be controlled by means of an extensive **controlling** process as a further field of action of the *Facebook* engagement (cf. Fig. 7.12) with regard to their effectiveness, in order to recognize possibilities of optimization at as early a stage as possible and extensively. But how can it be determined in the end whether a *Facebook* engagement was worth it? In order to do this, a comparison must be made between the *Facebook* targets, the investments in the *Facebook* engagement, and the results achieved based on the following *Facebook* **KPIs**, which are also relevant for other social media:

- **Reach**, measured by the gross and net reach as well as the first- and second-degree reach
- **Interaction** (conversational exchange), measured by the intensity of the communicative exchange
- **Engagement**, measured by the number of persons participating in a certain activity (e.g., competitions)
- **Sharing/distributing contents** (content amplification), measured by the number of persons passing on contents provided (e.g., shares)
- **Likes/rating of contents** (content appreciation), measured by the number of Likes generated
- **Mood** (sentiment), measured by the share of positive statements (in relation to neutral or negative statements)
- **Token**, measured as the number of own fans who have granted permission to access data

For the sake of structural "cleanliness," it must be pointed out that interaction, sharing, and rating are a specific type of "engagement." These KPIs can be declared as absolute values. Greater validity and easier comparability are provided by percent values, however. The challenge comprises the integration of these KPIs in a **social dashboard** in order to always keep an eye on the relevant developments.

> **Think Box**
> - How consistently are the fields of action of the *Facebook* engagements handled in my company?
> - Have we set up a community management system?
> - Who is responsible for the elaboration of the content strategy?
> - How long a term is this content strategy directed towards?
> - Has this responsibility been defined across the company or does each division work for itself?
> - How consistently is a promotion strategy elaborated and implemented—across online and offline media?
> - Has the responsibility for controlling been clearly defined and backed up by personnel and budgets?

(continued)

- Which *Facebook* KPIs are applied in my company?
- How regularly are the defined KPIs collected and put in relation to targets and budgets?
- Where can the overall responsibility for the *Facebook* engagement be found?

Various aspects of a **content strategy** with different **forms of engagement** will be demonstrated in the following. News in the fans' newsfeed can, on the one hand, get attention by addressing interesting topics followed by **seeding** (as in "seeding topics of conversation"). This way, an attempt can be made to initiate effective communication and such that is particularly aimed at smaller target groups. This is the case, for example, with TV programs which strive for an intensification of communication with the viewers via *Facebook*. This is how *Jersey Shore* (*MTV*) acts, for example. On the one hand, rather spontaneous forwardings can be strived for to cover a greater reach. The corporate objective here is frequently to involve **customers as brand ambassadors**. In this way, the corporate communication is extended by the users themselves and—ideally with positive contents—a greater reach is covered.

Posts represent important **triggers for various forms of collaboration** with fans. Fans can be sent off on an expedition together, develop and test products and services, elaborate slogans or entire advertising campaigns, or be integrated in other value added phases of the company (cf. Fig. 4.9). This **user-generated content** represents a particularly important result of a *Facebook* engagement with the acquired comments, ratings, recommendations, and ideas. **Product developments together with** *Facebook* **fans** have already been carried out in various industries such as cosmetics, textile, and tourism. However, companies should always reserve the "last word" for themselves when integrating users in the creative process, in order to avoid unpleasant surprises. When *Henkel* let the users vote on which layout a *Pril* dishwashing detergent bottle should be given, the majority voted in favor of "chicken fragrance—really tastes of chicken!" (cf. also article Ziems).

In a model competition by the mail-order house *Otto*, a man dressed as a woman was voted the winner! And at a photo competition, a photo with the visual slogan "Fuck U!" was liked the most, a fact that was not quite the intention of the initiators. For this reason, companies should clearly define the **rules of the game of a social media engagement** in advance, to which they can withdraw if necessary according to the motto: we reserve the right to make the final decision! The users should therefore never be able to act alone. In fact, it must be ensured that the company is the authority to make final decisions. The users must be informed of this restriction in advance.

A particularly interesting field of application for crowdsourcing was devised by *von Ahns*, the coinventor of the **CAPTCHA code**. This acronym stands for Completely **A**utomated **P**ublic **T**uring test to tell **C**omputers and **H**umans **A**part.

It is a tool that is able to determine online whether a computer or a human is accessing an online interface and thus acts as a SPAM protection mechanism.

With the system called **RECAPTCHA**, the Internet user now has to enter two words. With the first word, the system recognizes the solution and can thus tell whether a human or a computer is at work. The second word originates from a scanned book to be digitalized. When the user enters this second word, he makes a **contribution towards the digitalization** of the complete works, because a scan is being converted into a digital data set. In the process, several people carry out this digitalization process in order to achieve a correct result by cross-validating several entries. By doing so, words that the computer does not recognize can be safely digitalized by a crowdsourcing approach. The Internet has at the same time the **function of a conveyor belt**, which divides time-consuming processes into easily achievable steps. This way a great result can be achieved. About 100 million words are digitalized per day with this concept (cf. Heller, 2012, p. 73). The accomplishment of another large project, namely, the translation of the 6th *Harry Potter* book into German, was mastered back in 2005 via crowdsourcing. Within 48 h of the launch of the English version, the first German translation was available—even if with significant weaknesses. All in all, there are no limits to the creativity of such applications!

In the meantime, various **crowdsourcing platforms** have been developed to benefit from the intelligence of the mass. One example is *Mechanical Turk*, a project by *amazon*. This turns to inventors, developers, or other creative souls to motivate them—in return for payment—to cooperate. In parallel, companies are requested to post their tasks here in order to invite the mass to a creative cooperation (cf. amazonmechanical turk).

Galaxy Zoo is the basis for a large project (cf. galaxyzoo.org). This is about nothing less than the classification of galaxies on photos, which can be managed much faster by the input of more than 100,000 helpers than if only a few researchers were entrusted with it.

Frequently there are **winners of crowdsourcing** on both sides. On the one hand, companies can make use of the unlimited creativity of the Internet community to rapidly come up with innovations in a cost-efficient way and to receive input for the **innovation race**. This is reinforced by an ever greater pressure to innovate and reduce R&D budgets with shortened life cycles of new products at the same time. On the other hand, the techies, creative minds, and inventors frequently caught up in anonymity are finally finding attentive listeners who are seizing ideas turning them into marketable products. By doing so, under certain circumstances, important target groups may also be more committed to the company or brand. In this context, in our opinion, we need to speak of **customer-generated innovations**.

McDonald's appeals to its customers via *Facebook* to create their own burgers. Likewise, *Nike* and *adidas* customers are involved in the design process. Here, customers co-decide the design, the colors, etc., in order to create more individualized products. A professional implementation of **idea management** can also be found at *Dell* with *ideastorm.com*, which was brought to life in February 2007. Here, the customers are urged to actively cooperate and promised attractive rewards. After the customers have registered, they can write articles, go up or down in ranking by means of voting, and also comment on or rate articles by other users. More than 21,000 ideas have been submitted to *ideastorm.com*, which were commented on about 100,000 times and voted on 750,000 times. Roughly 600 ideas were already implemented (cf. IdeaStorm, 2014). The aim in the first instance was to offer customers a medium to express their requirements of new products or services and to turn them into reality. The winners are invited every year to a live "Storm Session" at *Dell World* where a jury votes on the ideas and, in parallel, be discussed by users on the website ideastorm.com (cf. *IdeaStorm* of *Dell*).

Tchibo—a German retailer company—also offers a convincing example of this with its Internet platform *TchiboIdeas*, where customers are invited to develop new products (cf. article *Tchibo Ideas*).

Article by Guest Author *Wolfgang Merkle*
The Idea Platform *Tchibo Ideas*: Background and Success Factors

Tchibo is a company in which the direct dialog with its customers is not only defined in writing in the **company DNS** as an essential behavior codex but is also actually practiced in the daily dialog in the shops. For it is only possible to constantly improve the range via the regular exchange with the consumers; it is only the direct, unfiltered feedback that can win the impulses so important for the optimization of the own performance. As a result, it was important to *Tchibo* consequently to aspire the **dialog with its customers and fans** even in the world of the Internet.

An additional incentive when establishing the **Internet platform** *Tchibo Ideas* was the aspect that *Tchibo*—in contrast to many other trading companies—develops its products itself and has its strategic suppliers produce them exclusively. In order to put this into effect, *Tchibo* has its **own product development process**, where trend scouts and product managers permanently work on developing and realizing new and desirous *Tchibo* worlds.

Tchibo Ideas was thus designed as an **online forum** to, on the one hand, understand the real needs of the customers even better. On the other hand, the interactive product and design development was also to be stimulated directly from the everyday problems discussed there via **crowdsourcing**. In this way, a community was born in which **clever product innovations** are developed from the discussions about individual ideas, some of which are turned into reality as concrete products—from a chopping board with an integrated collecting pan over the car handbag holder to a saddle cover firmly attached to a bicycle. These are all direct solutions to problems expressed by other users beforehand.

A key challenge when establishing such a **crowdsourcing platform** is first and foremost the establishment of the corresponding **community**. With *Tchibo*, consumers were looked for to describe their everyday problems and at the same time designers, too, who are happy to work on suitable solutions. The customers used their own weekly appearing media and the branches. Targeted

cooperations were concluded **with selected design colleges** to generate the idea competence.

For the permanent success of such a **crowdsourcing platform,** four key **factors for success** can be described:

Credibility
- Only if products and ideas from the platform are later turned into reality and available for sale are the members also ready in the long run to discuss their ideas on *Tchibo Ideas*.

Transparency
- The members want to know precisely what happens to the individual ideas. Only honest behavior towards the users in public discussions (comments, workshops, etc.) and in the realization of concrete projects (upcoming contracts, implementations, etc.) creates the trust necessary to keep such a community going in the long run.

Fairness
- Anybody who develops a completely thought-through product concept that is sold later in reality participates in the proceeds via a license payment for each product sold. Furthermore, the inventor of the respective product is named in all product communication.

Customer proximity
- Crowdsourcing is a tool which, above all, unfolds its entire potential due to the fact that the participants work on issues they are themselves interested in. Via open dialog and the specific possibility of directly influencing company processes, a company becomes touchable and shows that even individual opinions or precise problems of the customers really are important. At the same time, such a platform must be run in a very individual way. It requires a personal approach where the editorial team has to develop a good sense of the members.

Dr. Wolfgang Merkle, (Director Corporate Marketing, Tchibo, Hamburg)

However, this **process of customer-generated innovations** also involves risk. New ideas, insights, and solutions can be "ground down" and trimmed to the mainstream by the pressure of conformity of the mass. Some individual participants can frequently not afford to swim against this current. For this reason, one shouldn't pass on one's own intelligence to the mass—either as a private individual or as a company. At the same time, what *Steve Jobs* once basically said applies: "We don't ask the customers what they want. After all, the customers don't even know what is possible" (cf. Lashinsky, 2012). This is a good match to a statement by *Henry Ford*: "If I had asked my customers what they wanted, they would have said, a faster horse!" *Facebook* can, however, also be used as a **service platform** to provide customers with help and advice.

> **Think Box**
> - How consistently do we make use of the seeding of exciting contents on *Facebook*?
> - Are attempts made to consistently turn (satisfied) customers into brand ambassadors?
> - To what extent are attempts made to have user-generated content elaborated via crowdsourcing?
> - Are there guidelines—communicated towards the outside—for crowdsourcing to protect us from surprises?
> - Have we ever checked whether we can have internal tasks handled by crowdsourcing platforms?
> - Is it possibly productive to set up own crowdsourcing platforms?
> - Does *Facebook* offer interesting approaches to improve the quality of our service?
> - Who is predestined for the processing of these tasks?

A further interesting field of application of *Facebook* is described by the term **F-commerce** (also called *Facebook* commerce). This refers to a subset of social commerce. **Social commerce**—also called **recommendation retailing** or **social shopping**—is initially a specific development of e-commerce. E-commerce gets the "social components" by the active involvement of other users in the purchasing process. In this way, for example, their comments and performance ratings flow into the purchasing decisions as characteristics of the **ZMOT** (Zero Moment of Truth) and/or in real time via **online communication** with friends. The tools used in the process are called **social software. F-commerce** itself refers to the direct **selling of products and services through** *Facebook.*

The integration of shop functionalities takes place on *Facebook* via the so-called iframes. The use of these facilitates is, at the same time, the acquisition of large amounts of information about *Facebook* users. An interesting application **of F-commerce** is provided by *Facebook Connect* using the example of *amazon*. On the one hand, it facilitates the use of the *Facebook* **Interest Graph,** with the recommendation engine of *amazon*, to take the user's preferences into consideration which he has not yet revealed to *amazon* itself. By doing so, recommendations can be generated more personally and thus probably also more relevantly. On the other hand, an analysis of the *Facebook* **Social Graph** allows the network of friends and their information to be considered when generating shopping recommendations. In this combination, the information about friends' birthdays can be enriched by corresponding gift suggestions. The requirement for this access is the **permission** of the *Facebook* user. This token is obtained via a corresponding request page.

This is where the **currency CCCT** already introduced comes into action: a **contact** refers to a gift as **content** in the **context** of the birthday. If this relationship is built on **trust**, many users will grant permission and follow corresponding suggestions and make purchases. And if the **want button** is introduced on *Facebook*, such functions will appreciate!

But not only large providers such as *amazon* have the possibility to give their customers an **individualized purchasing experience**. There are various providers that also enable smaller companies to operate **F-commerce** by **integrating shop solutions** in the *Facebook* fan page:

Payvment (cf. ecwid.com/payvment)
- Payvment offers an e-commerce solution for *Facebook* that enables the integration of an online shop interface. At the same time, the application allows you to manage the shop itself as well as the offers presented in it and the sales generated via a corresponding dashboard.

BigCommerce (cf. bigcommerce.com)
- This e-commerce application enables you to integrate a shopping trolley function on the *Facebook* page. It makes it possible to look through product catalogs on *Facebook* and purchase directly on *Facebook*.

ShopTab (cf. shoptab.net)
- This *Facebook* application supports various shop functions (and thus various currencies).

How the application of *Facebook* can be used to **promote the sales of fan articles** is demonstrated by the example of *FC Barcelona* with their 54 million *Facebook* fans (cf. the following Baudis, 2012). In an intensive cooperation between *FC Barcelona* and *MicroStrategy,* a personalized **e-commerce channel** was set up based on *Facebook*. It also comprises a native *Facebook* application and a mobile app. One essential objective was to establish another **sales channel for e-ticketing and merchandising**. The *FCB Alert* was developed as a free *Facebook* app. This promises personalized and interactive experiences with the world famous football club. By providing **exclusive content**, such as match results, social games, photos, and contests, the fans are regularly animated to take part. This is meant to increase customer loyalty.

By means of this application, *FC Barcelona* have a technology platform "away from the field" to tie their worldwide *Facebook*-fan community even closer to them. *FCB Alert* opens up the following **possibilities of use**:

- **Interactive functions**, such as games and fan surveys (e.g., Who was the player of the day? Who was the player of the year? Down to questions such as: Are the gold-colored or green football boots better for our team?)
- **Access to exciting contents**, such as videos and other multimedia contents that can be shared with others

- **Access to news** about the team and the affiliated organizations which are pooled from websites, *Facebook* pages, *Twitter,* and blogs (e.g., *Tumblr*) to one place that is continuously updated
- **View of all** *Facebook* **events**, compiled by the club
- **Exclusive offers and advertising campaigns** of the team and the players
- **Purchase of goods and kit** of *FC Barcelona* via a *Facebook* store front

Via the *Alert* **platform** *FC Barcelona* can reach their *Facebook* fans from all over the world. The offer of contents can be created for various groups or individuals depending on their interest and/or demographic and geographical distribution. For this purpose, the app uses the user data available on *Facebook*—to the extent the users have granted their permission to do so. **Cloud-based social intelligence technology** allows *FC Barcelona* extensive analytical insights into their fan community. These include demographic and psychographic profiles as well as frequently changing information via check-ins and updates. Campaigns for special segments can be built up on these. By means of this, the user receives highly personalized and exclusive offers based on the data he has made available.

Yet there are also interesting applications available for the **traditional retail trade**. Together with the clothes manufacturer *GUESS* and *Tilly's Surf & Skate Clothing*, an innovative concept has been elaborated. These retailers can access their customers' *Facebook* accounts by using the **mobile commerce app** called *Alert* by *MicroStrategy* and link them up with their own loyalty program for the four million customers. The goal among others was to increase customer retention, generate customer insights, and increase sales (cf. Wohlfarth-Bottermann, 2013b).

Alert provides *GUESS* and *Tilly's* with a personalized and transparent channel with transaction possibilities for **social commerce**. If the fans have downloaded *Alert,* they receive personal messages, events, and targeted offers and vouchers of "their" brand, which are attuned to the social graph data on *Facebook*. *Alert* thus closes the circle between fan interests, customer segmentation, communication campaigns, and social commerce. This results in a **sustainable database for investments in social media**. The cloud-based social intelligence already presented in the *FCB* case enables *Guess* and *Tilly's* insights into their own fan structure. If a *Guess* fan, who has liked the appropriate *Facebook* page, goes shopping in London, he may receive the following messages on his smart phone: "Just around the corner, there is a shop with *Guess* jeans on offer. Click here to get a voucher ensuring you a 25 % discount"—certainly a more than suitable offer for a *Guess* fan.

Users of the *Alert* app can request their loyalty points and premium status, view their purchase history, and scan barcodes to get more product information. In addition, shops can be localized by mobile. And of course, purchases can also be made directly via the *Guess* online shop. What makes the app so special is that users have to log in via their *Facebook* account (social log-in), in order to be able to use the app. As soon as they have done this, the user must grant permission for the collection of *Facebook* profile data such as sex, age, family status, likes, check-ins, etc. and the utilization of it by companies.

In this way, *Guess* obtains a more precise picture of its customers. Internal CRM information such as name, date of birth, and address can correspondingly be enriched by level of income, level of education, or personal preferences such as music, cine films, hobbies, etc. This, in turn, facilitates a more in-depth and segmented approach to the customer. Via the app, *Guess* offers the customers not only personalized offers but also exclusive content matching his individual profile. The relevant added value and service achieved by this increases the readiness to download the app and grant the necessary permissions.

Tilly's, one of the fastest-growing retailers for surfer and sports clothes in the USA, faced the challenge of optimizing stocks without really knowing the customers. In addition, the company recognized that an integrated mobile strategy would provide an additional sales channel. Together with *MicroStrategy*, *Tilly's* developed the **mobile sales channel**. On top of that, *Facebook* was accessed **as CRM/loyalty database** in order to reach the customers with targeted and personalized offers. Besides the personalized offers and messages, the customers can also send vouchers to friends, collect bonus points, localize stores, and save them within the app as favorites. Besides that, receipts can be retrieved in the wallet and the barcode scanner can be used in the stores to obtain more product information and make direct purchases via *PayPal* (cf. Wohlfarth-Bottermann, 2013b).

This makes it obvious to what extent companies have become integrated in the *Facebook* environment with a direct objective of selling. The **relevance for the users** is also decisive here for success in this **ecosystem.** The examples demonstrated emphasize the fact that relevant contents and functions from a users' point of view are actually being offered—and with a great amount of convenience at that.

> **Think Box**
> - How significant can F-commerce become for my company?
> - Have we ever checked which starting points there are for us here?
> - Do we have the potential—alone or in cooperation—to develop comparable concepts such as the *Alert* examples?
> - Who could answer these questions in my company?

Alongside the applications demonstrated here, *Facebook* provides companies with a variety of **statistics and means of analysis**, in order to be able to monitor the use of *Facebook* offers. With a *Facebook* engagement, one should also be aware of the **risks**. Anybody who starts up on *Facebook* with no aim in sight and without the provision of necessary resources will almost imperatively be shipwrecked. And one more thing is important: there are no guarantees that one's own engagement in *Facebook* will automatically be successful!

▶ **Food for Thought!** **If you want a guarantee, then buy a toaster**—don't enter social media! Top management should also be told this!

In any case, it is important that we companies take great consideration of the *Facebook* **guidelines**. Breaches can be punished by exclusion from the world of *Facebook*, against which only a few "appeals" can be made. In the end, *Facebook* behaves in the same way as a diva who should best be treated with kid gloves. And not to be neglected are also the risks associated with the issues of **data protection** and **copyrights**.

Quick Wins

- _____
- _____
- _____
- _____
- _____
- _____
- _____
- _____
- _____
- _____
- _____
- _____
- _____
- _____
- _____
- _____
- _____
- _____
- _____
- _____
- _____
- _____
- _____
- _____
- _____
- _____
- _____
- _____
- _____
- _____

Why Marketing Is Becoming a Service

The **traditional understanding of service by companies** is as follows. The senior management responsible for service developed the so-called **Service Level Agreements**. These can be valid for their own staff or for external service providers. They stipulated, for example, on which days a customer service center is available (e.g., from 9.00 a.m. to 6 p.m., Monday to Friday; often explicitly not at weekends or on public holidays) and what the hours of business are. In addition, it was defined for the agents in the customer service centers, for example, how long a call had to last and which text blocks are to be used for customer correspondence. Whether this motivation corresponded with the needs of the customers or their customer value (cf. Chap. 5) was frequently not brought up or not sufficiently dealt with. The result of this was a situation as can be seen in Fig. 8.1. The **company** sees itself **as a conductor**—and the customers accept the services the way they are offered—or they simply don't.

Today, however, companies see themselves forced to offer an **ever-increasing range of services**, from which the customer can chose when and where he would like to avail himself of them. If nothing else, this trend is driven by the audience-grabbing **assertion of services** via social media (cf. Chap. 7). Here we can see that the allocation of roles between companies and customers is turning around to a greater extent. All of a sudden, customers are becoming conductors holding the conductor's baton and requesting service. In this case we can speak of a downright development of a **service cafeteria system** that makes one thing possible: **service of choice!** Here the customer becomes the **master of service** (cf. Fig. 8.2). Then the customer decides which services he wishes to request where and when. With that, the "where" not only covers stationary and virtual shops but also designates the actual place where the customer wishes to receive information, asks questions, and makes purchasing decisions. This can be in the subway, in the ranks of a football stadium, in the *Lufthansa* lounge, or in the bedroom. **Everywhere goes**! And companies are well-advised to focus on these new expectations of the customers!

The **development towards a master of service** leads to a self-service or crowdservice accompanying traditional customer service to cover the expectations

Fig. 8.1 Status quo of service rendering in many companies (*Source* Author illustration, data source: clipart)

Fig. 8.2 Customer becomes a "master of service" (*Source* Author illustration, data source: clipart)

of service. The possibility of involving customers in their own service rendering via **crowdservice** is interesting. To do this, we can set up platforms where customers help other customers (cf. Peppers & Rogers, 2011, p. 218).

Get Satisfaction is a platform helping companies to build authentic relations with their customers. *Get Satisfaction* is a customer engagement platform and powers 70,000 active customer communities—hosting more than 35 million consumers each month. They can use the *Get Satisfaction's* network to connect with each other to ask questions, share ideas, report problems, and engage with the brands and companies they care about. Hereby companies can engage with their customers anywhere they are: their own website, *Facebook*, via search, on a stationary or mobile device (cf. GetSatisfaction, 2014). *Kellogg's* is an interesting example.

> **Think Box**
> - Is there a development towards a "master of service" in our sector too?
> - Have our competitors already responded to it?
> - What possible consequences for the service we offer are becoming evident?
> - Are there appropriate concepts for crowdservicing for my company to reduce service costs and/or increase customer retention?
> - Who is responsible for answering these questions?

An important aspect that plays a significant role in this context is called **gamification**. This describes the use of elements that are appropriate for games—yet incorporated here in a non-game context. These can be scores achieved, ranking lists, and awards the user can obtain. The hands-on elements are used to increase the motivation of the participants when, for example, monotonous or longer lasting tasks need to be accomplished. The displaying of friends on *Facebook* or the Likes received for certain photos or posts make use of exactly these hands-on elements to involve the user in a permanent relationship. For these named scales have no upper limits and they promote and encourage continuous engagement—to "look good" in a social context!

A convincing example of this is provided by the so-called **social check-in services**. Concepts such as *Foursquare* and *Facebook Places* allow user to check in at physical places. To do this, the application accesses the current location via the GPS of the mobile device being used. At *GetGlue* they can check in to shows, movies, or sports that they consume. By doing so, they inform their friends of their current location or activity. In addition to this, there are the following possibilities, depending on the concept respectively used:

- If a certain location has not yet been listed, then the user can enter it for the first time.
- The location being visited can be directly communicated to one's own network via *Facebook* and *Twitter,* for example.
- It is possible to receive tips from one's friends and strangers who have already been there and have recommended or warned.
- At the same time, the user can write ratings that are in turn available to other users of these places. Once again, the do ut des principle applies: I "out" myself and my location as well as my preferences or ratings and, in return, I can access recommendations and ratings by others.
- In addition, *Foursquare* offers the possibility of earning "laurels" for frequently checking in to certain places. The person who checks in to a certain place the most frequently receives from *Foursquare* the rank of a mayor.
- Incentives of the location owners can be combined with such ranks if it is, for example, a café or a restaurant.

- Apart from that, the rank increases the "social significance"—in some cases depending on the type of location visited!

Such **hands-on possibilities** now need to be combined with the **increased expectations of service**. Are the customers, in many cases, not even ready to pay for personal service? Isn't there the possibility for some companies to offer a **concierge service for those customers** who are willing to invest more?

What are the key **demands of service** to motivate the customers to pay for them?

- **Convenience in access**, e.g., also by mobile, and at all times of the day or night.
- **Relevance of contents**; this means a targeted playout of service offers.
- **Engaging the offer**, i.e., the users should be prompted to participate.

On the way to a completely **new type of service quality,** we can draw on the image of a **butler**: only due to the fact that he is so close to "his master" based on his spatial and contentual level and is also informed about all (secret) preferences is he able to serve us the *Earl Grey* tea at 11.30 a.m. precisely with two pieces of candy—or the favorite Shiraz wine in the matching *Riedel* glass with the preferred finger food in the evening. Anybody who is ready to reveal more about himself at least has the chance of an extraordinary service experience if this readiness is also supported by the necessary purchasing power (cf. Chap. 6). Or more concisely, **service and privacy are two sides of the same coin**. Only if the customer says what he wants is that what he also gets. Here is an ice cream parlor as an example. The salesperson asks "Which ice cream would you like?" And the customer replies "That's none of your business, that's private." The customer will then maybe get a nice mixed ice cream but not what he really wants.

> **Remember Box** No data. No exclusive services!

In 2012, *KLM* started the project **social seating**—called *Meet and Seat*. *KLM* can identify the ideal seatmates on flights for travelers who allowed *KLM* to access the *Facebook* or *LinkedIn* profiles of their passenger. If someone wanted to learn Mandarin, for example, *KLM* could arrange for finding a seatmate for the 8-h flight to New York who not only speaks Mandarin but also loves teaching it (cf. KLM, 2014).

Or imagine the following situation. You arrive with a *Lufthansa* flight in London on Wednesday evening. As your business appointment is not until the next day, you can do what you wish in the evening. How would you find it if, shortly after arrival, you check in at a **social check-in service** such as *Facebook Places* and a few seconds later, you receive the following message from an **entertainment platform**: "Welcome to London, Mr. Land. Do you have any plans for this evening? We know you like going to classical music concerts but also that you have a liking for pop music. We can offer you a seat in the *Royal Albert Hall* at a classical concert with *Daniel Barenboim*, or a ticket for an *Adele* concert. What may we do for you? Or

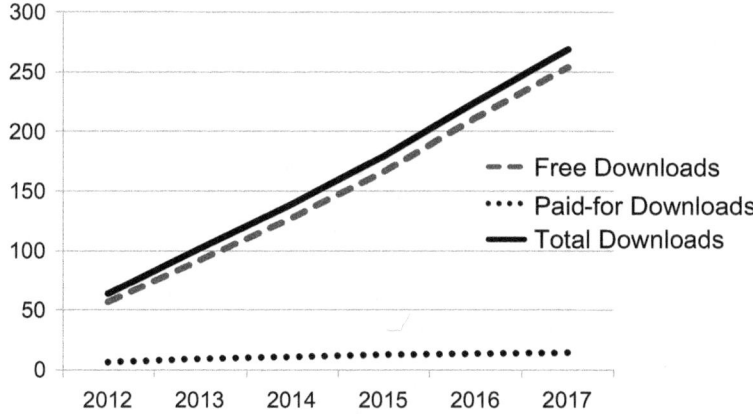

Fig. 8.3 Mobile app store downloads, worldwide, 2010–2016—in billions (*Source* Author illustration, data source: Gartner, 2013b)

would you prefer to go back to your favorite restaurant in Soho?" If we have allowed the provider access to our *Facebook* data, such offers are no longer dreams of the future. And the readiness to pay more for such a service than only the ticket price must very high.

A key driver behind such services will be the increased development and the **increasing use of apps,** because they alleviate the access to very special services without the need for elaborate search processes on the Internet. The fact that such applications are becoming increasingly more popular is shown by the increased number of users downloading mobile apps in Fig. 8.3. This is why an **app economy** is rightly so spoken of. In our opinion, it is thus essential for all companies to do an extensive check for their own company as to whether relevant contents and especially (paid) services can be offered via apps—either alone or in cooperation with partners. If the development of the smart service terminal described in Chap. 1 is continued, more and more companies will consider themselves forced to provide not only online contents but also mobile via apps.

Such apps are also increasingly finding their way into **smart TVs**. Specialized service providers are conquering the living rooms more and more with their video-on-demand offers. A requirement for this triumphal procession is the cooperation with the TV manufacturers. In the long run, TV will become an app that "happens" to run on a screen that we used to call a television! In sum, the individual screens will increasingly merge, fed from the cloud so that the user can switch between various devices without leaving "his" ecosystem! The race is on. And in this case, too, the one who wins will be the one ensuring the most convincing holistic customer service.

> **Think Box**
> - What effects does the development towards an app economy have on my company?
> - Do we have or can we set up contents and/or services for mobile access?
> - Is it worth it for my company to develop apps ourselves or should we do this in cooperation with diligent partners?
> - Do we focus rather on content or on relationship—or do we have an ideal combination of both to offer?
> - How can my company exploit the trend towards "gamification"?
> - Of what significance are check-in services to us?
> - Where do we see the possibility of regarding, developing, pricing, and selling "marketing as a service"?
> - Who can deal with these issues?

In addition, completely new scopes for design are emerging. For a long time now, studies have been demonstrating the consequences of when high performers—be it in a private or professional environment—are permanently distracted from their "actual" tasks by emails (at the PC but also via *Blackberry*, *iPhone*, and *iPad* or any other mobile device at the same time), text messages, advertising banners, and calls. According to a study by the *University of California*, staff members in large companies are barely able to concentrate on one task for 11 min at a time before there is a distraction. It takes the employees 25 min to return to the original task. The **time losses** and **inefficiencies** associated with such a way of working were estimated by *Basex* for the USA at about 600 billion US $ per year. Now imagine: A study in 2013 found out that 25 % of a student sample uses the smart phone more than 2 h daily. On overage they activate the smart phone 80 times a day—i.e., every 12 min on average in the daytime! The emergence of the disorder **ADT** (Attention Deficit Trait), commonly known as **CMS** (Constant Multitasking Craziness), is associated with this (cf. Claus, 2014, p. 7).

A study by *Stanford University* revealed in addition that **multitaskers** cannot master a task better than people who work on one task after the other. On the contrary, multitaskers tend towards distraction, are more susceptible to diversions and interruptions, and work more inaccurately (cf. Wiedlich, 2010, p. 1). Thus, **multitasking** does not constitute a convincing solution for coping with the increasing number of tasks. This equally applies to professional and private everyday life.

There are primarily two approaches to these challenges for those concerned. On the one hand, the **self-organization** of each individual is required to protect oneself against the **information overkill**. The synonym for the information overkill is the information ping which, for example, notifies of an incoming email or text message. This information overkill is frequently combined with an **activity overkill** in line with the motto Every action as in an email, a message, etc., requires a direct reaction—"mercilessly" geared towards the now and real-time demands of the Internet. However, it would be a more than worthwhile approach to "grant" oneself

8 Why Marketing Is Becoming a Service

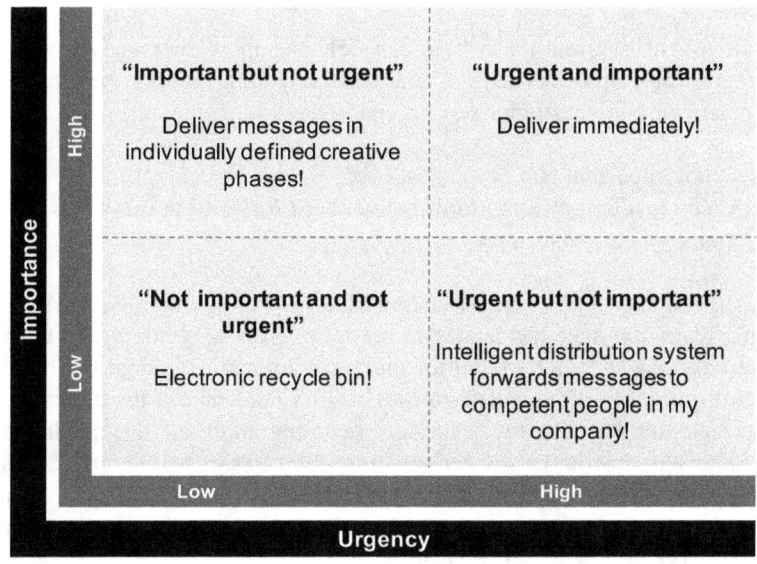

Fig. 8.4 Meta-master—classification according to priority and relevance (*Source* Author illustration)

communication-free time frames—oriented towards the principles of **time management**—which conduce to concentrated working. In some cases, such approaches are discussed under the buzzword **slow email movement**. The tasks associated with this are first to be tackled by us an individual but also organizationally for our company.

On the other hand—and this is where exciting spheres of activity result for us both as a company and as customer as well as a service provider—in the future, there will be **meta-masters**. Their task merely comprises filtering out those messages with the greatest relevance from the large number of messages each individual is bombarded by in an intelligent, dynamic way and in real time. Thus, like many of us today use a pop-up filter and a virus scanner, meta-masters will sort out the information we are bombarded by into the following categories according to **priorities** and **relevance** (cf. Fig. 8.4):

Category "urgent and important"
- Deliver immediately! This includes, for example, information about meetings, work from the boss, or invitations by the mother-in-law.

Category "important but not urgent"
- Such messages can be delivered in the individually defined creative phases to experience profound handling (e.g., a suggestion for a product innovation or a report about the customer complaints from the week before).

Category "urgent but not important"
- An intelligent distribution system can delegate these messages to competent colleagues or employees (e.g., a standard complaint that can be handled with competence by the customer service center).

Category "not important and not urgent"
- Electronic recycle bin (incl. information about forwarding messages, meaningless "thank-you emails," etc.).

Where necessary, these **meta-masters** also send automatic **acknowledgments** informing that a message has landed in the recycle bin or sending will be delayed until the next day in order to control the expectations of feedback. This type of **M2M communication (machine-to-machine)** would take a great strain off the original recipient but "bestow" further incoming mail on the sender. In this instance, the sender would also need a meta-master to sort out incoming messages. The necessity of actually finding such solutions results from the fact that the amount of data—as already stated—will also increase dramatically in the future; yet productive work still needs to be ensured.

M2M communication for taking the strain off the user will still take place; however, when cars automatically make an appointment for a vehicle inspection, refrigerators analyze shortages or expired best-by-dates and automatically cater for supplies from the regular online shop. Mobile phones automatically top themselves up on the Internet; health monitors make an appointment for the annual health check and recommend, where applicable, paying for it yourself to benefit from a no-claims bonus. Estimates speak of 26 billion devices being connected to the Internet in 2020 (cf. Gartner, 2013c). A large field of application for "marketing as a service"! Customers will be ready to pay for such meta-masters but also for **decision support systems**. In the end, customers save time and money in this way (respectively depending on the personal opportunity costs).

Supplementary to the abovementioned classification of incoming mail—on whatever channel—additional personal requirements can also be handled in a competent manner. When inquiring about a restaurant—using **location-based services** as an example—only those restaurants are automatically displayed—based on personal preferences as to how far one wishes to travel. In addition, it is checked whether the corresponding restaurants also have free tables on the desired day, have good customer feedback, and whether they accommodate the user's preferences as to "affordable," "romantic," or "modern." Of course, as well as that, the personal preferences in taste (Italian, French, Chinese, Indian, and German style) are taken into consideration by the selection. It can also be taken into consideration that—in spite of a high affinity to French cuisine—one does not appreciate going to French restaurant twice within 7 days. The chosen restaurant can be directly booked via an interface to a **reservation system**. In this case it also applies that only such offers can be found that are also available online. After visiting the restaurant, a rating is requested to check the "fit" between user and offer. This results in further corporate fields of action for **marketing as a service**.

> **Food for Thought** The challenge for companies and providers of service platforms is not the generating of an app for everyday tasks. The task is rather the **development of a master app** revealing the quality of an all-rounder which takes care of the users to an ever-increasing extent in the **own ecosystem**!

It becomes apparent: marketing becomes a service when a holistic perspective of **customer experience management** is adopted. The customer is seen and cared for as a whole and involved information-wise via the pre-sales, sales, and post-sales phase. And with each further contact, further information depicting the preferences even better is gathered so that the "next service" can be rendered even more personalized!

In this case, too, the previously described **law of disproportionality of information** (cf. Chap. 6) applies: only if customers are ready to grant us access to their *Facebook* data, for example, are we able to render such individual services. With that, it becomes clear that the "**currency trust**" is becoming more important.

Where are we heading? *Bernd Stahl* of the network specialists *Nash Technologies* worded it as follows: "In the future, we will see nothing at all of the communication. One can access the intelligence of the Net from anywhere. There is no need to be concerned about special devices any more, the choice of services, the network or service provider. I no longer need to enter a destination via telephone numbers, IP addresses or links. It will all be taken over by intelligent, semantic networks. The relevance of the enquiry is automatically recognized, the enquiry is split up into single parts, sent to various destinations and back comes the desired service or finished product" (Sohn, 2012). And this vision has partially become reality.

> The brave new world is waiting for us!

And we should try to profit from this trend towards paid services—that is, "marketing as a service."

Think Box
- What necessities to act are there in my company so that my own employees are not drowned by the information flow?
- What approaches can we use to master the data flood?
- Does my company have the potential to set up "sellable" services for the provision of information to people?
- Do we see potential to participate in the development of meta-masters?
- Who could we give the corresponding research work to in my company?

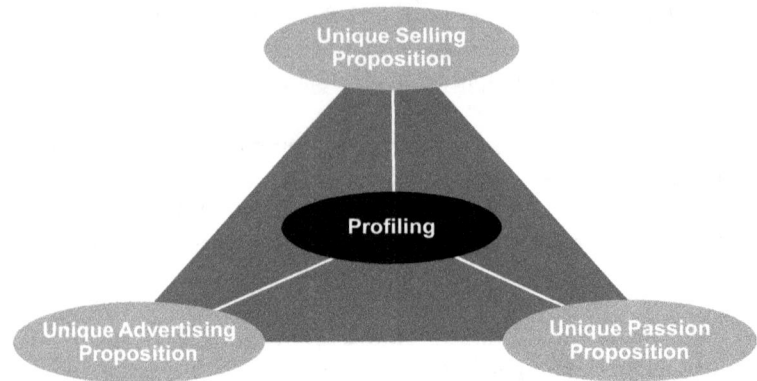

Fig. 8.5 Approaches for reaching uniqueness (*Source* Author illustration)

In order to achieve an extraordinary service quality in such a way, a new concept for achieving **uniqueness on the market** is moving into the spotlight: the **unique passion proposition** (UPP; cf. Fig. 8.5). Yet what differentiates this UPP from USP and UAP? A **unique selling proposition** (USP) is based on "objective" and thus provable facts which some companies keep secret. This is the case with the likes of the recipe of *Coca-Cola* and *Underberg* or the original recipe of a sauce at *Kentucky Fried Chicken*. Others apply for patent protection (like in the pharmaceutical industry, engineering, or electrotechnology) to safeguard a long-term competitive advantage that can be effectively applied in marketing. The **unique advertising proposition** (UAP), which aspires an island position of the brand by means of promotional staging and in contrast to an "original" element of use is frequently more difficult to copy, must be differentiated from this USP. The advertising message by the body spray *Axe*, "*Axe—the fragrance that provokes women*" or "*Axe—and you'll never walk alone*," thus "only" conduces the setup of a UAP for differentiation in the competitor environment without ever actually honoring this promised performance in reality. This also goes for the advertising promise "*Red Bull gives you wings*"—something that is not really fulfilled after the stratosphere jump by *Felix Baumgartner*!

The **unique passion proposition** is about enhancing the objective, the service offer, albeit is a brand, a concrete product, or a service in the eyes of the customers by making the **passion** of the people acting behind it visible and tangible. Maybe it is even possible to align an entire company in a "passion-driven" way. The differentiation from the USP is achieved, even though with the UPP, no "facts and figures" can be cited for documenting the superiority. It is more about the "spirit" behind a range of services. In this respect, a UPP is also significantly more than a UAP, which is merely created by communication, without accessing objectively provable facts.

If this **spirit** becomes visible for the potential customer or customer, especially in the service area, then his purchasing decision can be positively influenced by this—according to the motto "If the employees work flat out for their company, their brand, their product, then it has to be something!" In this way, insecurity in the decision-making process to purchase can be reduced and trust can be established. Yet a UPP is not achieved until it becomes obvious **in the eyes of the target group** that **passionate actions** are behind a company, a brand, or a service which can be substantiated in various dimensions:

- The passion to render an excellent service for the customer (if necessary, "24/7")
- The passion for having the best product on the market and for continuously developing it further
- The passion for "going the extra mile" for the customer
- The passion for never resting on one's laurels but to be spurred on to new success by success

With that, it is important that this passion is "real" and not artificial, just because the employer wishes it to be. It is thus about the passion to achieve **service excellence** for the entire company (cf. Fig. 8.6). Many companies will only be successful in the future if they **gear their organization towards passion** and at the same time fill out all of the service fields presented in the service excellence turbine with the same amount of passion.

In the process it will be revealed that even those companies whose marketing strategies or offers are less innovative than those of the competitors can be more successful. A requirement for this is that the strategic concepts are implemented across all company hierarchies and the partners involved and arrive at the customers as a **passion-driven organization.**

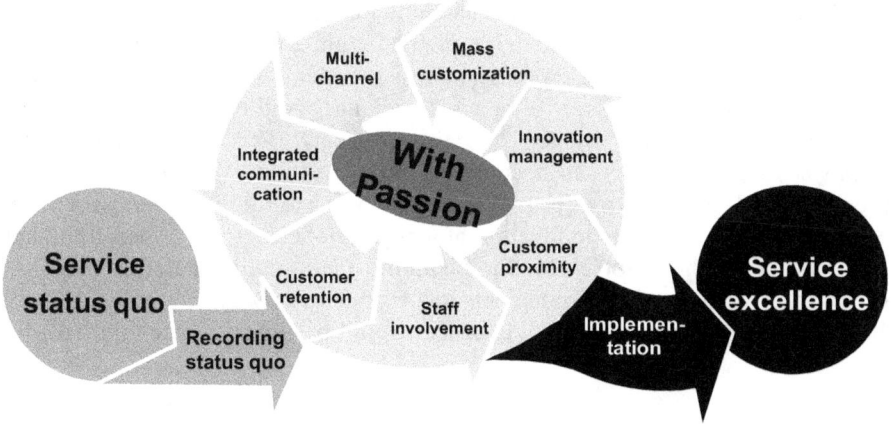

Fig. 8.6 Passion-driven service excellence turbine (*Source* Author illustration)

▶ **Remember Box** The only thing that cannot be copied in the long run are the relationships a company, and in particular its management and its employees, establishes with customers. **Service excellence** can make a significant contribution towards this.

▶ **Food for Thought** **Enthusiasm is good fuel, but unfortunately it burns too fast.**
Albert Schweitzer

In face of these challenges, isn't it high time to promote the "customer retention ladder" to the "**customer enthusiasm ladder**" or the "**customer trust ladder**"—with regard not only to the name but also the actual doings? Wouldn't that be a much more powerful statement—equally directed towards the outside and the inside? For being committed to "one's" company is something the least customers strive for! The target is to become a real lovebrand.

If that is to become the name of the game, there are further **functional descriptions** that show a significant about-turn in the focus of the companies. For instance, the highly awarded seminar hotel *Schindlerhof* in Nuremberg has defined a management function on senior management level which is quite simply called "**cordiality officer.**" Their main responsibilities comprise conveying to the customer "a service experience distinguished by cordiality" across all hierarchies and process levels. A senior manager entrusted with such a task could also bear the grand title **chief experience manager** or **chief customer officer**. This would express that it is about the creation of a self-contained, valuable customer experience—and again across the entirety of the previously discussed customer trust points.

A requirement for this kind of service experience is "active listening." Thus a consequence of this is that—in the case of *Dell*—the first **chief listening officer** (CLO) was employed on C level. It is essential that his function is not restricted to "mere listening." The CLO is rather equipped with extensive power to shape in order to also prompt change processes across the hierarchies and corporate management as they are necessary after his "listening." And of course, a CLO also needs many active "co-listeners" in the company. For this reason, 17,000 employees have already been accordingly trained at *Dell*: 2,500 of them were even qualified as **social media professionals** who are allowed to represent the company to the outside in the function of a press relations officer. Apart from that, *Dell* has set up three **Social Media Listening Command Centers** worldwide, which analyze about 27,000 statements in 14 languages every day 24/7 (cf. Buck, 2013). A title such as chief listening officer would be—in the context provided—

much more comprehensive than a management board section called **chief social media officer** or **chief digital officer**, which, in some cases, only shed light on partial aspects and lack the holistic view (cf. short article *Dell*).

Article by Guest Author *Michael Buck*

Effective Customer Service and Authentic Dialogs with the Customer in the Web-2.0 Era: The Example of *Dell*

It was great shock when *Jeff Jarvis*, in his capacity as US-American journalist, professor, and blogger, in 2006 delivered his **crushing judgment** about *Dell*: "Dell sucks. Dell lies. Put that in your Google and smoke it." Yet *Dell* acted. And today, *Jarvis'* words sound completely different: "In the age of customers empowered by blogs and social media, Dell has leapt from worst to first." So what happened in the meantime?

Even if the blogger scene was still limited in 2006 and the influence on the representative opinion on the web still had limited reach, *Dell* realized relatively fast that the company had to become prepared for an **appropriate dialog** in these new channels. The involuntarily received feedback from the blogger *Jarvis* opened up the opportunity for the company to recognize potential **weak points in communication** at an early stage and to remove them step by step in the years to come. New innovative approaches emerged for a sustainable improvement in the company.

Dell accepted this **chance as a challenge** and knew how to make use of the revolutionary potential for innovation in it. The results speak for themselves: **Products** and **services** are becoming **measurably better, customer loyalty is increasing,** and the **brand is gaining trust**. Through the innovative use of **social media applications** and the **intelligent use of existing social networks** such as *Facebook, Twitter,* etc., *Dell* has taken on a pioneer position in social marketing. By means of the **dialog** with their own customers and potential customers of the brand as well as through **structured listening,** the company receives valuable feedback—in real time and regarding many different issues and products. In this way, an understanding for the customers and their purchasing decisions develops—"the why behind the buy."

Two key aspects were in the focus of the use of social media in the company right from the beginning. And without dispute, they still form the key components even today: **customer contact** and **recommendation marketing**. Two other key aspects have recently emerged: the **advanced customer involvement** and an extensive understanding about why customers buy what they buy. Yet the process must not end with **listening**.

Dell became aware of the advantages of social media very early. The **support and promotion by the senior management** (first management) played a significant role in this. The widespread use of digital media at *Dell* already leads today to a radical change in marketing, PR, human resources, sales, service, and communication with the staff in the company. Today, *Dell* already has trained more than 17,000 employees worldwide in the field of social media and thus implements a far-reaching transformation into an even more customer-oriented global brand.

Social media transform companies and have great **influence on corporate culture and organization.** This can easily be identified in the example of *Dell*. The brand is developed further by customers; the customers have greater influence on products and on how the brands are perceived by other customers.

Inside and outside the company, the following applies: Social media do not need to be controlled; they merely need to be used by companies in a meaningful way. Customers expect a **new quality of communication** from the providers on the market. They will be demanding this direct connection to the company more and more. **Recommendation marketing** has reached a new dimension in which the customer is not only a provider of feedback but is downright invited to act as consultant, diplomat, and brand ambassador for the company via the **social exchange**. The moment the company not only gives the customer the opportunity to express himself but also listens to him and implements what it has heard to become better for the customers, the right path has been taken.

Michael Buck, (former manager of the worldwide online marketing at *Dell,* today strategic corporate consultant for the digital corporate transformation)

Until now, it has been demonstrated what interesting approaches the increased consideration of customer preferences has for marketing. However, the **social revolution** described in Chap. 2 as well as the **social CRM** involves an entirely different effect which ever-growing circles are becoming aware of little by little. Due to the fact that—especially online—more and more information is available about us and our preferences, we are increasingly becoming the **prisoners of our own preferences**. This phenomenon is described by the term **filter bubble** (cf. Fig. 8.7) coined by *Eli Pariser* (2011). The online providers on the Internet

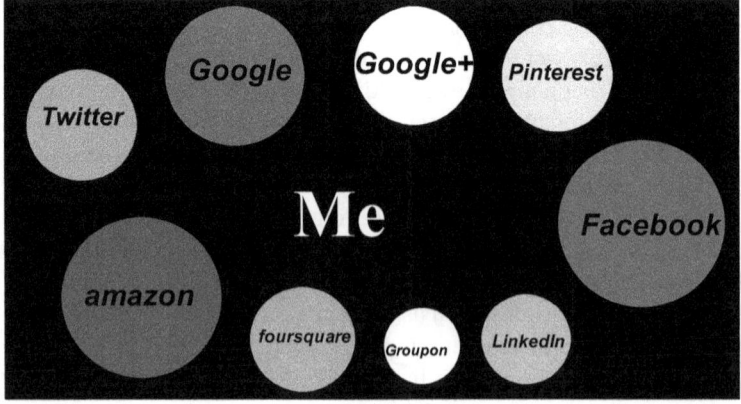

Fig. 8.7 Filter bubble—imprisoned in the network of our preferences (*Source* Author illustration)

are trying harder to get access to our preferences in order to submit the best-suited—because relevant—information and offers. If we occupy ourselves with the information provided over a longer period—and this in turn is recorded—our already identified preferences will be confirmed. This procedure is repeated if we, for example, accept the offer presented by *amazon*. The consequence: we receive increasingly "more of the same."

This process involves a continuous **restriction to the access to information and offers**, initially unnoticed by the users. For *Google, Facebook, YouTube,* and the likes are increasingly only submitting those offers where we reveal the greatest "probability of purchasing." The effect of the filter bubble thus increasingly restricts the way we see the world! And the view of the world presented to us differs from that of other users—with diverging preferences. In this way, the opinions we once formed are becoming increasingly manifest because we receive fewer alternative points of view and offers.

The effects of the **relevance-based offers** are known to us from *amazon*. Anybody who has once demonstrated interest in a book of *John Grisham* is constantly sent information about new publications by him. Whereby this form of the **individualization of offers** and the underlying mechanisms of a **shopping basket analysis** still remain recognizable, we cannot recognize the same effects with *Google*. Here the results presented by *Google* change depending on the **click and surf behavior on the hit pages** of the search engine. And even on *Facebook* we influence from which users we receive status reports through our user behavior. The mechanics behind this are simply called: **targeting**—or more precisely—**behavioral targeting** or **predictive behavioral targeting** (cf. Kreutzer, 2014 for more detail). The entirety of these procedures fully analyzes our **digital footprint** (incl. our digital shadow). When the **want button** is introduced on *Facebook*, the offers will be even more customized because concrete wishes of users can be relied on.

These analyses lead to one phenomenon that can be called the **endless content loop**. For—based on the previous insights—only similar offers are presented because they reveal the **highest purchasing expectations**! If we paint in black and white, the following consequence may eventuate: "We die the virtual death of predictability" (Meckel, 2011, p. 94).

The filter bubble represents a particular risk when—as has already frequently happened—governments influence the search engine operators so that they no longer make certain contents available. This is where a political filtering of information takes place, which reduces the "objective" view of the world.

> ▶ **Food for Thought** The **filter bubble** increasingly blocks us from obtaining information without hindrance which—merely seemingly—is made accessible just as easily or just as difficultly for all on the Internet. Which information we see, hear, or read online is thus no longer determined by us but by the algorithms of the large information providers. With that, these providers determine our view of the world, the companies, and their offers—still more or

less unnoticed by the public. The consequence: **we see the world through a filter which we have set up ourselves.** And this filter is determined by everything about the world we have been interested in.

And which **consequences of the filter bubble** and in particular the social filter contained within do we need to consider when communicating? The social filter leads to the forwarding of information we have received being successful if it is also relevant for friends. That means nothing other than us having to conduct **communication via the cushion**—analog to pool—in order to reach the target people!

Quick Wins

- _____
- _____
- _____
- _____
- _____
- _____
- _____
- _____
- _____
- _____
- _____
- _____
- _____
- _____
- _____
- _____
- _____
- _____
- _____
- _____
- _____
- _____
- _____
- _____
- _____

9 The Necessity of Change Management: Why Our Traditional Communication and Organizational Structures Are Becoming Obsolete

The exploitation of the potentials of the upcoming changes will not be successful without a comprehensive **change management** to master die digital transformation. The position we are currently at in this process can be seen in Fig. 9.1. Are we still the **"observers"** watching the "new stuff" with interest without already being real **"listeners"** who, for example, have set up a web monitoring system? Or do we fall into the **"analyst of changes"** category, with which a more profound investigation of the challenges defined by social media with regard to the own business model is involved? Or has there already been a **"piloting of first test projects"**—the necessary intermediate stage towards **"strategic and organizational anchoring"** of the questions about the social revolution? Or have we already achieved an **"active participation as day-to-day routine"** and adjusted our structures, processes, and range of services to the integration of the potentials of social media in a holistic way?

On the basis of this rough analysis, it is necessary to enter the **phases for opening up the social and digital potential** of a company. Figure 9.2 illustrates the steps where this process or the integration of social media can be developed. The **social media newcomers** stated in Chap. 1 are mainly to be found in **Step 1: Experimental phase** (also cf. Forster, 2012). This involves initiating first attempts to walk without real corporate commitment—frequently without the decided allocation of staff and financial resources. The entire activity rather takes place under the title "youth researches"—a fact that in part holds true, given the age! Guidelines for social media activities as well as appropriate monitoring are lacking. The **social media pioneers**, who have been dealing with various social media applications for somewhat longer, can be found in **Step 2: Setting up social media islands**. Here, first social media applications are started within the company and operated with limited use of staff and finances. An overall social media or digital strategy cannot be recognized in these approaches either; nevertheless, first guidelines are drawn up and monitoring tasks elaborated. The majority of the staff regards the company's own engagement as an "exotic without real potential."

210 9 The Necessity of Change Management: Why Our Traditional Communication...

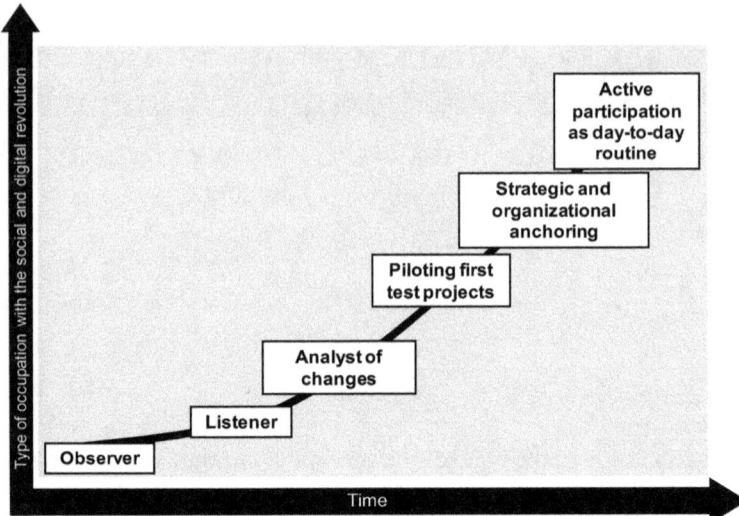

Fig. 9.1 What is my company's position while mastering the social and digital revolution? (*Source* Author illustration)

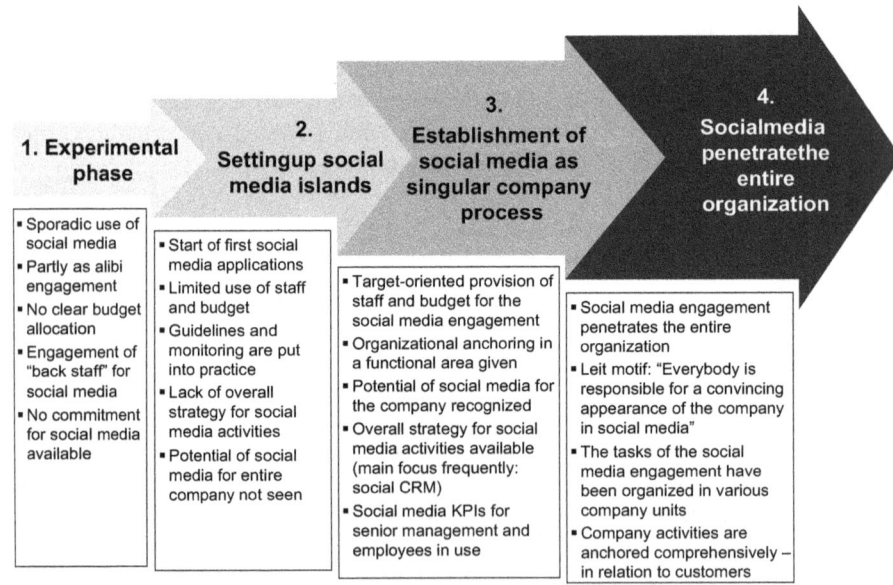

Fig. 9.2 Stages of development for exploiting the potentials of social media (*Source* Author illustration)

Some of the **social media pioneers** are already in the transition to **Step 3: Establishing social media as a singular corporate process** (cf. Fig. 9.2). In these companies, the great potential of social media and the digital transformation to safeguard and extend a company's own field of business were recognized and put into a functional shape regarding the organization. Staff and budget were made available—oriented towards the goals to be achieved. Marketing frequently took over the function of overall responsibility whereby a focus is frequently seen on social CRM. In order to guide senior management and employees, corresponding KPIs are used by means of which the commitment of top management becomes visible.

Step 4: Social media penetrates the entire organization represents the most comprehensive type of the organizational anchoring of social media marketing. Here, the cross-corporate engagement in social media penetrates the entire organization—as is the case with "market-oriented corporate management"—due to the marketing principles. At the same time, the activities in social media have given up their close connection to one functional area (frequently marketing) and penetrate the entire organization. With that, the guiding principle applies: "Everyone is responsible for the convincing image of the company in social media!"

It is comprehensible that the **need for change management** in the first three steps of this process is particularly pronounced. After all, it is necessary to extensively further develop the existing operational and organizational structure. In doing so, not only do information and process silos need to be forced open but also areas of responsibility that no longer fulfill the new requirements of the **social and digital era** must be changed.

> ▶ **Remember Box** The utilization of the chances of the social and digital revolution requires a **systematic change management**. Hence, social media engagement must always begin internally—first of all in the heads, then in structures and processes. Not until after that should social media engagement become visible to the outside!

Think Box
- Where is my company in the mastering of the social and digital revolution?
- Which phase of the development process for exploiting the potentials of social media is my company in?
- Have we ever determined the potential of social media marketing and the digitalization of the world for the entire company?
- Which drivers and constraints can be determined staff-wise in the entire organization?

(continued)

- Which hurdles can be determined from an operational and structural point of view?
- Have we honestly done our "homework" before becoming active in social media as a company?
- What is the position of the top management towards social media and the digitalization challenges?
- Are we already willing to think and act more in network structures and in projects instead of in static and stiff hierarchies?
- Can we succeed in speeding up the operational and structural organization to bring greater task orientation into the company?
- Who can become the "driver" for the necessary change management process?

In this book, a large number of measures which are necessary for **surviving the era of digital Darwinism** and to **master the digital transformation** have been demonstrated. The knowledge that companies have to adapt themselves, their business models, and brands to the new requirements is, however, only just starting to become accepted on the various levels of the company. Yet a strategic bottleneck remains: the **implementation**. How do we succeed in turning "strategy into action"?

Many brilliant concepts and strategies haven't made it from being on paper (or the digital equivalent) to being done and ended up as "ideas quasi locked up in the poison cupboard," never to see the light of day. What **hurdles** do the managers themselves who are working on the **implementation of digital strategies** see? Figure 9.3 gives a first impression. In first place—named by 81 % of the managers interviewed—are the **lacking competences** to accommodate the altered framework conditions. As was the case with many other waves of innovation, the **IT landscape that has evolved over time** represents the stumbling block in 43 % of the companies questioned. This technological gap is revealed with likewise 43 % with the **lacking competence to connect mobile platforms** to the company's own **ERP software**. These findings refer to the situation in Germany. But we have recognized—based on our research and additional discussions with CEOs and CMOs in many other countries—that the hurdles are very similar there, too.

Another interesting issue mentioned by 28 % of the managers is the **generation gap** which is documented especially in the openness towards social media and rather facilitates an evolution than a revolution. The key question is thus: Will the market and thus the customers as well as the competitors give the companies the time necessary for reorientation? The generation gap also seems to have had a negative impact on the **development of a digital concept** in general: in part, the digital concept has been elaborated too **sequentially** and is **not** aligned **holistically** enough. It is also remarkable that the **threat to established channels by virtual channels** is after all regarded as an "obstacle" by 9 %. This threat is in the meantime simply a fact in many sectors. And only those who courageously take on this

9 The Necessity of Change Management: Why Our Traditional Communication...

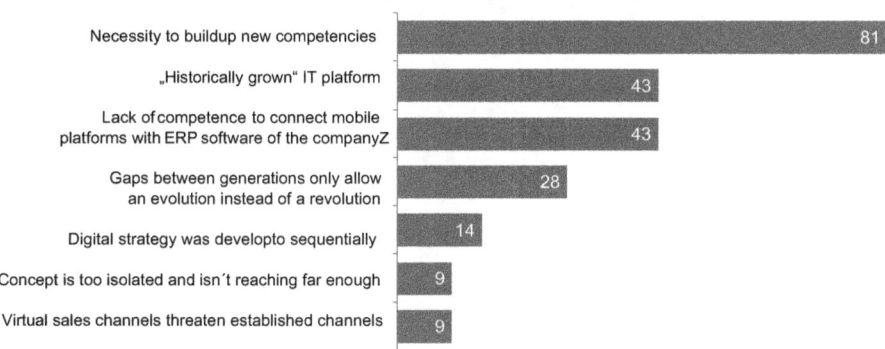

Fig. 9.3 Which aspects hinder the implementation of a digital strategy—in % (Germany, $n = 100$ managers, several answers possible) (*Source* Author illustration, data source: Camelot Management Consultants, 2012, p. 21)

challenge will survive! The consequences a hesitant approach can have can be read in the newspaper every day!

> **Remember Box** In order to survive digital Darwinism, an extensive **change management process** is needed. And it needs to begin top-down to be successful. At the same time one thing is certain: in the course of this process, important comfort zones need to be surrendered!

But what are the consequences of the necessities for change that have become clear in the previous chapters? Are the companies already working intensely on mastering them and using their **senior management and staff as strategic resources in the fight for competitive advantages**? Far wrong. Reality still is a rather "lousy mood" in far too many companies. You don't believe that? Convince yourself by taking a look at the annually published data by *Gallup*. *Gallup* (2013) distinguishes between three types of employees:

- **Engaged employees** work with passion and feel a profound connection to their company. They drive innovation and move the organization.
- **Not-engaged employees** were essentially "checked out." They are sleepwalking through their workday, putting time—but not energy or passion—to their work.
- **Actively disengaged employees** are not just unhappy at work. They are busy acting out their unhappiness. Every day, these employees undermine what their engaged coworkers accomplish.

The *Gallup Organization* conducted a study on the extent of the **engagement between employees and companies in the USA and Canada**. According to this study, **70 %** of the employees in the USA did **not feel any real engagement**

	Engaged	Not engaged	Actively disengaged
United States	30%	52%	18%
Canada	16%	70%	14%

Fig. 9.4 Development of the engagement index of *Gallup* in the USA and Canada (*Source* Author illustration, data source: Gallup, 2013, p. 85)

towards their work: **52 % work to rule** and **18 %** have already **inwardly quitted**. With this, the share of employees with little or no engagement to their profession has reached an alarmingly high level (cf. Fig. 9.4). The share of employees in the USA revealing a **high emotional engagement** to their job or work environment is merely approx. **30 %**. The situation in Canada is even worse. Only 16 % are engaged—84 % are not engaged or actively disengaged.

If one compares the results of the **Engagement Index** in the USA with all other countries being analyzed, then it is demonstrated that a 30 % share of employees with a high emotional engagement is a strong above-average value. The **average percentage of engaged employees** in 142 countries is only 13 %! The corresponding figures of engaged employees are for the UK, 17 %; Germany, 15 %; and Italy, 14 %. In countries such as France (9 %), Turkey (7 %), as well as China (6 %), a high emotional engagement is almost fully lacking (cf. Gallup, 2013, pp. 111–113).

If one analyzes the results of *Gallup* from 2008 to 2012 as per Fig. 9.5, then it becomes obvious that this is not a temporary problem but a longer lasting process. The figures about the **farewell into inner emigration** have remained at a high level for years and have done so despite a difficult economic state at times. However, the knowledge about the extent of the inner termination of the emotional disengagement of the employees to the own company has hardly initiated conceptual processes as was the case, for example, with customer retention. In companies, things other than enthusing employee and senior management performance are still being focused. As already shown in Chap. 8, this is an imperative requirement for **reaching a UPP** (unique passion proposition). The result that has become visible in the data presented reveals a **resistance by the employees**. With that, their level of performance remains significantly and sustainably below the potential available. *Gallup* estimates that for the USA, active disengagement costs US $450 to 550 billion per year. In Germany, that figure ranges from US $151 to 186 billion per year. In the UK, actively disengaged employees cost the country between US $83 and 112 billion per year (cf. Gallup, 2013, p. 7). There is huge room for improvement!

There are significant effects connected with the different degrees of engagement. It is obvious what approaches companies could pursue to overcome the **engagement gap**. At the same time it becomes evident that the fulfillment of the **basic needs** ("material and tools are available," "know what is expected") in a company does not reveal any extreme differences with the three employee groups without, with little, or with high emotional engagement. Things look different when it comes to the **management support** experienced. Here are the following aspects relevant:

9 The Necessity of Change Management: Why Our Traditional Communication...

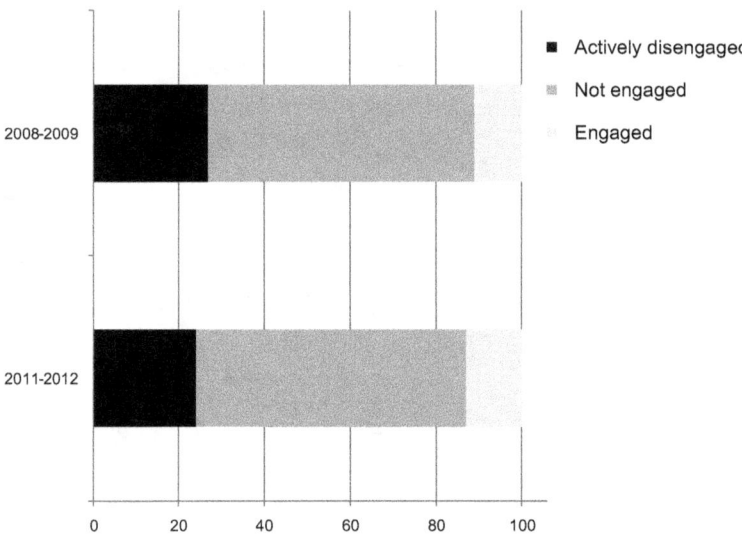

Fig. 9.5 Average engagement index of *Gallup* among the employed population in 142 countries worldwide (*Source* Author illustration, data source: Gallup, 2013, p. 14)

"personal development," "be regarded as a human being," "appreciation," and "do what I can best." The employees with a high emotional engagement feel they are supported by their company significantly more than the other two groups. Even **teamwork** is rated significantly better by the emotionally highly engaged group. Teamwork is defined by the following aspects: "good friend at work," "colleagues engage for quality," "corporate targets are known," and "our opinion counts." After all, the emotionally engaged employees see better possibilities of personal **growth**, i.e., "learning and development" and "progress." From the point of view of the employees without emotional engagement, this practically does not take place. Each company is required to analyze the status of the emotional engagement in order to derive appropriate action for improvement.

The effects of **resistance** are manifold. Based on a comparison between top- and bottom-quartile teams, significant median differences could be identified. In the top-quartile teams the absenteeism is 37 % lower.

Gallup conducted a **meta-analysis** using 263 research studies across 192 organizations in 49 industries and 34 countries. *Gallup* calculated the business/work-unit-level relationship between employee engagement and performance outcomes that the organization supplied. This analysis confirmed the well-established **connection between employee engagement and performance outcomes** (cf. Fig. 9.6). *Gallup* studied the differences in performance between engaged and actively disengaged business/work units and found that those scoring in the top half on employee engagement nearly doubled their odds of success compared with those in the bottom half. Median differences between top-quartile and bottom-quartile units were 10 % in **customer ratings**, 22 % in **profitability**,

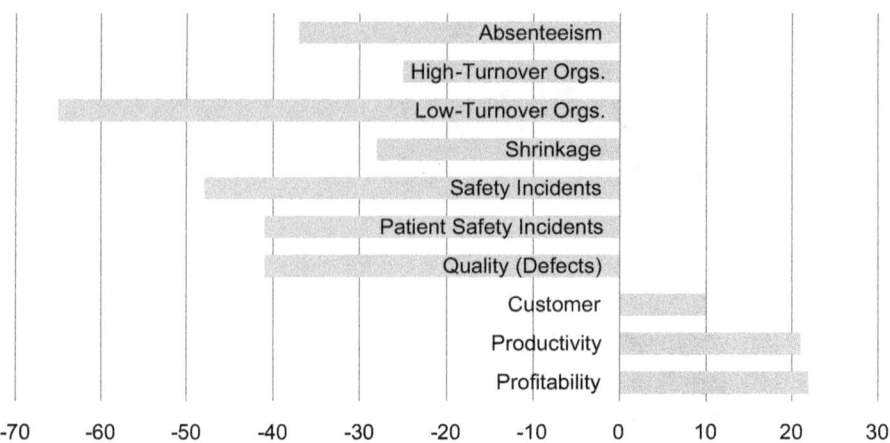

Fig. 9.6 Engagement's effect on key performance indicators—in % (*Source* Author illustration, data source: Gallup, 2013, p. 22)

21 % in **productivity**, 25 % in high-turnover organizations, 65 % in low-turnover organizations, 48 % in **safety incidents**, 28 % in **shrinkage**, 37 % in **absenteeism**, 41 % in **patient safety incidents**, and 41 % in **quality** (defects). This meta-analysis verified once again that employee engagement relates to each of the nine performance outcomes studied. It also shows that the strong correlations between engagement and the nine outcomes studied are highly consistent across different organizations from diverse industries and regions of the world (cf. Gallup, 2013, p. 21).

If companies strive for a strategic further development and differentiation in competition in the era of digital Darwinism to achieve the digital transformation, in our opinion, **employees and senior management** must no longer be neglected as the (most) important **success factor**. They must fill the strategic orientation and the values based on it with life. At the same time, due to the increasing significance of services, employees will be generating an increasingly larger part of the company's value creation because the established industrial nations are increasingly becoming **service communities**. This means nothing other than employees and senior management being given increasingly greater significance as central resources because they are involved much more intensively in the **value creation process with and for the customer** (buzzword "marketing as a service" in Chap. 8).

A result of this is first the necessity that the staff reveals both customer and sales orientation (cf. Fig. 9.7). **Customer orientation** merely resorting to "making customers happy" is not enough for profit-oriented companies. This customer orientation must be brought into balance with **sales orientation**. For this reason, all measures taken in the context of staffing policy need to be analyzed to see whether they contribute towards result-oriented goals of the company (cf. Chap. 5).

Another factor that reinforces the relevance of an extensive consideration of the potentials of the own senior management and employees is the increasing necessity

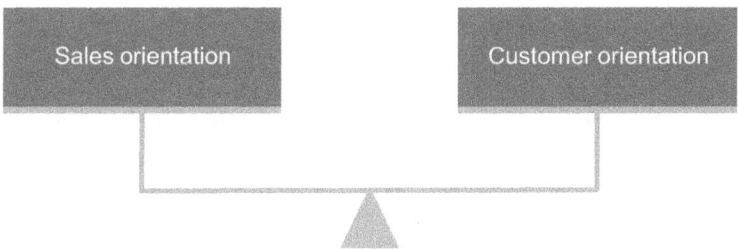

Fig. 9.7 Ensuring balance between sales and customer orientation (*Source* Author illustration)

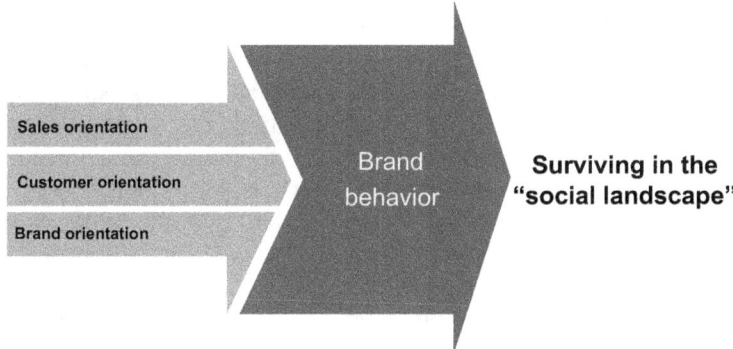

Fig. 9.8 Establishing brand behavior by ensuring a balance between sales, customer, and brand orientation (*Source* Author illustration)

to differentiate with offers about the **quality of services** becoming increasingly similar in competition. The relevance of a **unique passion proposition** was already demonstrated in Chap. 8. The issue is now to gear the organization to its provision because for certain companies a solid basis for a long-term competitive advantage can be achieved by means of a unique passion proposition by **focusing on the "passion" factor**. Yet at the same time, it is imperative to gear and thus channel the passion to be aroused to the company's brand promise and the respective offers. For this reason, customer and sales orientation need to be supplemented by **brand orientation** (cf. Fig. 9.8). It is this triad that first leads to the aspired **brand behavior**, behavior by senior management, and employees that contributes to the overall positioning of the company in the online and offline environment.

The relevance of such an approach is underlined by the **social currency** concept by *Vivaldi Partners Group* (cf. short article *Vivaldi Partners*). It becomes evident that the **social brand value** is given increasing significance in comparison to the **monetary brand value**. A high social brand value is in turn reflected in the monetary value of the brand.

Article by Guest Author *Erich Joachimsthaler*
Social Currency: Creating New Business Models

Ever since the industrial revolution and before, companies have created value by producing and selling things consumers needed or wanted. For many decades now, through cycles of economic prosperity and decline, serving generations of consumers, companies grew large, powerful, and rich; others disappeared.

Since those early days, **management wisdom** has it that companies can sustain growth and prosperity only by developing a **competitive advantage**. This helps to charge a premium price for products or services, and it helps fend off competitors. There are basically two sources of competitive advantage: cost leadership and differentiation advantage. **Cost leadership** can be typically achieved through major efforts and investments in efficiency improvements, better plants, better control of overheads, and efficient deployment of R&D. Companies can develop specific skills to exploit any cost advantages. For example, fashion apparel retailer *Zara's* ability to turn over product on the shelf in the now famous 15-days cycle gives it important time-to-market advantages that made it one of the largest brands in its category in the world today. **Differentiation** advantages can be achieved by focusing on quality, design, service, and a better emphasis on branding, brand advertising, and reputation.

The larger, the bigger, and more powerful a company, the more it can exploit cost leadership and differentiation advantages; the greater the volume of production, the greater will be the productivity in terms of working methods, labor efficiencies, and plant utilization; in other words, **size matters**. Companies with leading market shares in a category or sector benefit more than others. The more dominant the market share, the more they are in the position to create and extract value. This has been the logic of the industrial age forever.

Today, this logic is merely a half-truth, and in many industries and categories it is flat out false. *Nokia* had a dominant share in the global mobile phone market and just missed one turn towards smart phones. Today, *Nokia* struggles. *Best Buy* was the dominant electronic retailer in the USA that won with customers and put *Circuit City* out of business based on store location advantage that *Best Buy* had. Three years later, its retail store advantage feels like a millstone. It now fights survival against *amazon* and other online retailers. *Sony* had the strongest brand in the music and electronics business with tens of millions if not hundreds of millions of *Walkman* listening devices in consumers' possession by the year 2000. *Sony* clearly was the market leader, but this enormous size advantage did not help *Sony* a bit when *Apple*, a computer company, launched the *iPod*. As *Gary Hamel* said: "Incumbency never has been as worthless."

The Social Age

We believe that today, the industrial age, or era if you will, has been supplement by the *social age* where value is no longer just created by companies producing products and services or things. In this new age, value is created through the connections that individuals have with others. As *Nilofer Merchant* says: "If the industrial age is about *building* things, the social age is about *connecting* things, people and ideas." This is a **new source of competitive advantage**. Some examples include:

General Electric—*GE* has been a **front-runner in the social revolution** among large industrials. It has an enormous presence on all major social networks. It runs *MarkNet*, a social network that operates inside *GE*. Just recently, its chairman and CEO launched an entire new range of products and services that signify an entire **new source of value creation** which is about "The convergence of the global industrial system with the power of advanced computing, analytics, low-cost sensing and new levels of connectivity permitted by the Internet." *GE* is talking about the *Industrial Internet* and its most ambitious effort to date to leverage the **connectivity of things**—jet engines, trains, and power plants—and exploit the information and data resulting from connectivity.

Wikipedia—*Wikipedia* is a free online encyclopedia that was started by *Jimmy Wales* in 1999. Today, it contains more than 21 million articles and has over 470 million unique visitors. It is edited by more than 88,000 volunteer "editors." The site is a good example of the **power of the so-called wisdom of the crowd**. It has 18 million registered users who have made an average of about 20 edits per page. The advantage of this large community of editors cocreating *Wikipedia* was one of the reasons why the venerable, over 240-year-old *Encyclopedia Britannica* had discontinued its print edition in 2012.

Microsoft—*Microsoft* bought *Skype* for US $8.5 billion, even though *eBay* sold *Skype* a year earlier at a value of about US $2.75 billion to a group of investors. During the year, there wasn't a lot that had changed at *Skype*. Subscriber numbers stood at 663 million. The advantage of people texting, video chatting, sharing files, and talking to each other on this technology platform when combined with other services, such as *Microsoft Dynamics CRM*, *Bing*, *Yammer*, and *Office*, were so valuable for *Microsoft*, it paid nearly three times what *Skype* was deemed worth just a year earlier. *Microsoft* did not overpay, even though an outside observer that looks at the financials would rush to conclude. To *Microsoft*, it is a chance to reinvent how businesses around the world perform their daily workflow processes and how people connect.

These three examples illustrate that **value is created not in the thing itself but in the connections it fosters**. For *GE*, the data from these connections create enormous competitive advantage in delivering efficiencies and cost savings to clients. For *Wikipedia*, the information that people voluntarily share with others creates its product, a new business model and a disruptive insurgent for *Encyclopedia Britannia*. And for *Microsoft*, *Skype* represents a platform where

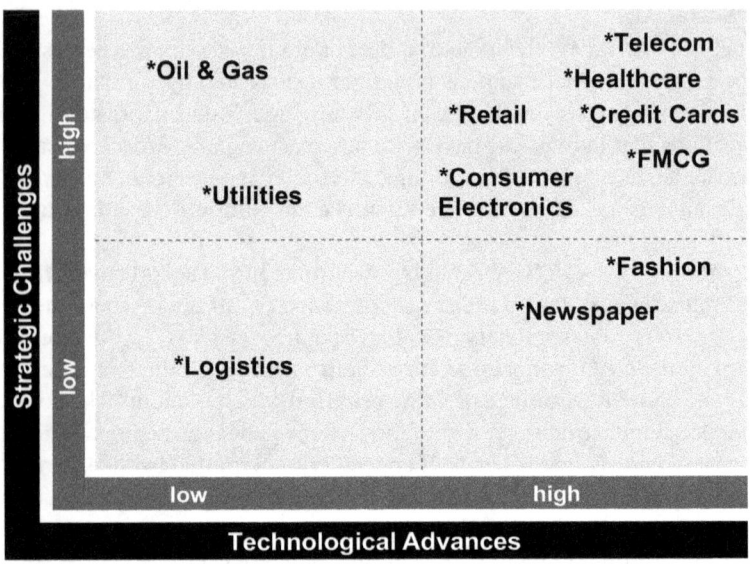

Fig. 9.9 Social impact evaluation matrix (*Source* Illustration by Joachimsthaler)

millions of people share information daily with others to manage their lives and their businesses.

Put another way, the **value of a company** by these three examples relates to the **relationships or connections** it can foster. Of course, this **new competitive advantage** does not replace typical advantages of cost or differentiation, but it certainly amplifies those advantages. The overall impact on industries could be seen in Fig. 9.9.

Social Currency

Not all connection and information are of equal value. Some are more valuable than others. Information that is created by people such as consumers or customers is of particular interest to businesses and marketers. Over the last few years, these types of information have grown disproportionally fast. A recent study by the IDC technology researcher showed that more than **75 % of the information in the digital universe is generated by individuals** today. And the total amount of information doubles every 2 years. On top of that, more information is generated about consumers through mobile devices, sensors, and other technologies.

This represents an enormously valuable amount of information generated by consumers or about them—not by companies, but by regular people who might be customers, friends, or even foes—around brands and businesses, liking them, following them, talking about them, and making connections with friends, family, and others. This information grows exponentially because customers and consumers are ever more empowered, enabled, and informed about technologies than before when it comes to engaging and connecting with brands and businesses. Because of the value of this information, we have developed the **social currency concept**. The concept measures the degree to which customers share a brand or information about a brand with others as social currency. With social currency, we attempt to understand how this information can be leveraged and exploited by companies to build strong brands and businesses. Today, in our opinion, companies do not excel in adapting to this changing market paradigm. At *Vivaldi*, we recently studied the **social behaviors of customers** and **how they connect with brands** online and create social currency through a survey of more than 5,000 customers in the USA, the UK, and Germany. We focused on more than 100 brands and businesses across 19 categories. We also studied several hundred social and digital initiatives by major brands in the USA and Europe. We attempted to answer two key questions:

- First, what are these **social behaviors of consumers** that underlie social currency, and does social currency create competitive advantage?
- Second, how well do brands and businesses today **exploit the information resulting from social currency** that are enabled through the new social and digital technologies?

Six Dimensions of Social Currency: The "Social Six"
Our research shows that there are **six different types of behaviors of consumers**: utility, information, conversation, advocacy, affiliation, and identity. These behaviors are not new consumer behaviors. Consumers and customers have engaged in these behaviors even before the Internet, but our research shows that today's digital technologies, particularly **social technologies, change the prominence and relevance of these behaviors** during purchase considerations:

- **Utility**—Exists when consumers derive new value from engaging with brands and other people. This value can range from giving simple discount such as when *The Gap* ran a deal with *Groupon* and sold over 441,000 coupons of US $50 for US $25 to providing consumers with a way to measure physical activity throughout the day using the *Nike Fuelband*. The *Fuelband* creates utility value because it motivates users by awarding points that helps them live by the numbers.
- **Information**—Exists when consumers receive from and share with other people valuable information about brands. *Fab.com,* for example, provides its 7 million members with information from their social network to help in making online purchases.
- **Conversation**—Exists when consumers talk about a brand or business to others. Think *Old Spice*. The brand has played on humor with its *Isaiah Mustafa* videos on *YouTube* and positioned itself against the enormously success *Axe Body Spray*. Just one vid: "The man your man could smell like" was viewed more than 43 million times. These views have been multiplied in countless conversations among consumers anywhere.

- **Advocacy**—Exists when consumers or customers actively promote or defend a brand or business to others. *Audi* targeted design, luxury, technology, and automotive influencers for its *A8* top of range model. It invited these influencers to a test drive who then spread the word about the new *A8* to their followers and fans. The *Audi* effort wasn't just a word-of-mouth program, it was a carefully orchestrated program to maximize advocacy.
- **Affiliation**—Exists when consumers connect and become a member of a community of people that is linked to a brand or business. *Philips*, the global technology firm, has built a large community on *LinkedIn*, called *Innovation in Light*. It is one of the largest B2B networks that deliver powerful community benefits such as informed discussions, chats, information, and connection.
- **Identity**—Exists when consumers express themselves to others in relationship to the brand. *Levi's Jeans* created an identity building effort called *Levi's Girl* by asking female consumers to create a video about themselves and explain why they are the best spokesperson for the brand, representing the *Levi's* values. Contestants could ask their friends and fans to vote which translated in self-expressive benefits.

Our research shows that across all brands and businesses, these **six dimensions drive consideration, purchase,** and **loyalty** when companies actively enable consumers along these behaviors. **Social currency** has a **significant impact on brand equity** (53 %) and **drives important brand equity dimensions** such as **quality** (26 %) and **loyalty** (28 %) and **key brand perceptions** such as **liking** (34 %), **trustworthiness** (35 %), and **authenticity** (33 %). There are huge differences across brands, category, and industry (see for report of the research results Vivaldi, 2012a).

Best Practices
The second question our research intended to answer was how well do brands and businesses today exploit the new social and digital technologies? We analyzed over 100 case studies from around the world (see Vivaldi, 2012b). Through this research, we learned that **most companies do not exploit well these new technologies to build competitive advantages and strong brands**. While a large percentage of firms monitor social conversations, set up communities, and launch social media initiatives to amplify communication campaigns, such efforts are often executed as if social or digital is merely another channel or customer touch point.

We found that decisions concerning social and digital are often disconnected from the decision-making processes of the marketing organization and, even worse, are misaligned with company strategy. Digital initiatives abound, of course. Every function from market research to customer service and communications and advertising to R&D includes a digital project or social initiative. But more often than not, these are siloed efforts that **lack the proper coordination and integration with the overall strategy**. The leading best practice is to set up a digital or social center of excellence that is managed separately or merely located within the communications function.

It is time for executives who wish to build strong brands and realize competitive advantages to begin to **rethink their business strategy** and how they are organized in order to execute in today's connected world. Such an effort must also encourage a closer look at the processes, systems, and policies of companies' systems. It must accelerate the way information is delivered to key decision-makers across organizational functions and geographies, so that consumer insights, trends, and key challenges for the brand are identified, analyzed, and acted upon in real time. It must enable and support strategic new processes and not merely pile more data on top of existing data.

A Case Study

As it is with any new technology, its impact across categories, sectors, or industries is not uniform. Of our 19 sectors studied, we believe the new digital and social technologies have a **significant impact** in the **fashion, apparel,** and **retail sectors**. Hence, it is instructive to analyze how social currency impacts company's and brands' activities and lead to entire new ways of doing business. From our research we chose two fashion retailers *Zara* and *Burberry* to describe some key emerging principles.

Zara has become one of the most successful fashion retailers. One dimension of its success is the company's capability to relentlessly optimizing the traditional fashion and apparel retail value chain. *Zara* takes just 4 weeks to turn an idea into merchandise and items spent 2 weeks on store shelves. Technically, this makes *Zara* one of the **fastest retailers**.

The company relies on business information and analytics, so-called Big Data, to manage inventory, logistics, and distribution. **Consumer preferences drive the value chain activities**. Through its success in optimizing the supply chain, and the entire value-generating activities, *Zara* has also become a popular subject of study by students in MBA programs and executives across all kinds of industries.

If we were to position the *Zara* success merely as a function of optimizing the supply chain, it would not properly reflect the competitive advantage this fashion retailer has developed. *Zara* has developed an **entire system of a set of activities** that are sequentially executed and logically aligned **to capture value** and **deliver a superior customer value**—from how *Zara* identifies consumer preferences to how *Zara* fashion apparel reaches the market. The modern term of such a system is called the **business model**. In our experience, it is extremely difficult to compete against such a well-developed business model. There is enormous competitive advantage.

Yet, there are developments in the fashion and retail sector that suggest that even *Zara's* successful business model can be upended. The key is **social currency**. Our research showed that several fashion retailers are working with the new digital and social technologies to integrate even more deeply consumers

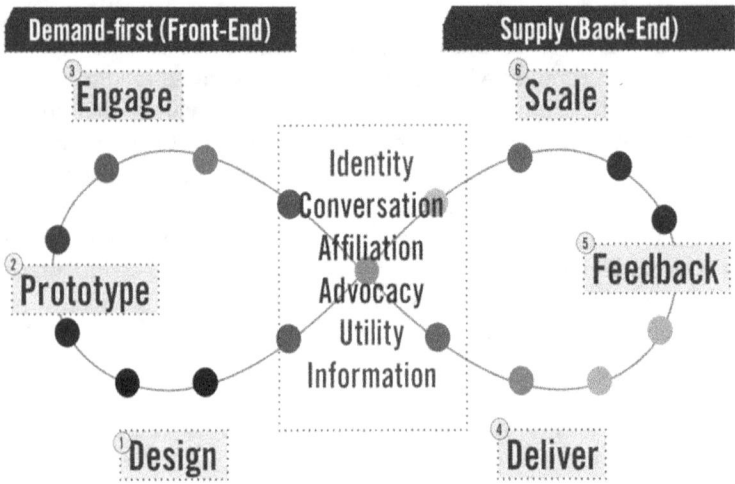

Fig. 9.10 The new demand-first value chain (*Source* Illustration by Joachimsthaler)

into the various activities of the value chain. Some even uproot the traditional logic of the linear value chain. A good example is the iconic British luxury brand *Burberry* that might well be on its way to build the next business model in fashion retailing.

Burberry is a much smaller fashion house concerning revenues. It operates still most of its business operations following the traditional value chain serving 700 stores worldwide with its products. In addition, *Burberry* also experiments with an entirely new business model by innovating in the front-end of the value chain. We call this model the **demand-first value chain** because it creates value in unique ways with an innovative demand-driven front-end phase first that reconfigures the traditional value chain in new ways, followed by a back-end phase of scaling the success model (cf. Fig. 9.10).

Reimaging the Traditional Value Chain
Imagine it would be possible to design only a small number of apparel items, manufacture the small batch, and test them on millions of shoppers. Imagine you could get the orders from millions of consumers and their credit card payments before you make any commitments to suppliers of the fabric, yarn, or other materials or logistics. If that were possible, you would produce only what you know you can sell, you would need no financing since you have already charged the consumer, you would never miss another product forecast, or miss a logistics plan. You would not need marketing or promotional efforts to move product off the shelf, no more clearance sales, no need for outlet stores to sell off excess inventory.

While such as business model is hard to imagine, *Burberry* has initiated some efforts that make some of these thoughts a reality. In order to build its fan base on *Facebook* (13.5 million versus *Zara's* 14.6 million) and create a massive *Twitter* following (*Burberry* has 1.25 million followers versus *Zara* which has 90,000 followers), it launched *TweetWalk* during its Spring/Summer 2012 collection. Fans who liked the brand on *Facebook* were invited to watch its fashion shows in the major cities like Paris or Milan from the comfort of their desks or whether the computer or tablet was. The initiative was hugely successful because fans could watch the rehearsal tapes of the shows 1 h before models actually hit the runway. This **created massive conversations** on *Twitter* as fans used the opportunity to express their opinion and shared their liking or disliking for the collection even before or while the actual fashion show unfolded.

Burberry has intensified its presence on social networks. On *Instagram* it counts 525,000 followers while *Zara* has 4,500. On *YouTube*, it has 41,000 subscribers and 16 million views versus *Zara* with 6,000 subscribers and 1.5 million views.

With its **runway to reality effort**, it made its videos shoppable for the Fall/Winter collection. While shoppable videos is now already relatively common, it created for customers of *Burberry* a chance to click on the video of a model featuring the new collection, pick a size, add to the cart, and have it pay for with a deduction from the previously stored credit card. Shoppers have a choice to get the merchandise delivered to their home via express mail within 2 weeks or to pick it up at one of *Burberry's* many store locations; that then creates another shopping opportunity.

Using Social Currency to Create a New Business Model
It is our opinion that *Burberry* is at the forefront of building the next new business model for the fashion and the apparel industry. From a **value chain perspective**, its front-end and customer-facing end has three major steps: design, prototype, and engage with fans and consumers. This demand-first cycle changes how consumers are involved in **three value-generating activities**. There is instant feedback from customers about actual demand for particular products at every stage—design, prototyping, and communications or engagement. The possibilities of cycling through this process are infinite. It is the possibility to run as many front-end cycles. While *Zara's* optimized value chain reaches a shelf turnover of a month or in the extreme even 15 days, *Zara's* upper limit for turnover can be even shorter. In the extreme, the front-end cycle could be run daily receiving valuable information about demand for particular designs instantly. Once the shopper has made the payment, the second cycle or back cycle starts with delivering the product to the store or home, receiving feedback about preferences, which provides additional business information about scaling the business.

In order to fully understand the **value of this potentially new business model**, it is helpful to learn **how it creates social currency**:

- **Utility**—For those who are fans of *Burberry*, this model provides a new channel and even more convenient way to shop for the latest fashion anytime and anywhere. Fans can influence the collection through voting; in a way, they are participating in the business of *Burberry* and have a say about it.

- **Information**—Fans of *Burberry* have information about the latest collection before the fashion magazines have reviewed it. The time lapse from engaging with the *Burberry* brand reduces since a consumer can review fashions anytime, not only when during regular store visits at the beginning or end of a season.
- **Conversation**—Consumers feedback information to *Burberry* and engage in conversations about anything that really matters to them. It is not about customer service but about brand engagement and brand building, creating a natural conversation with consumers about the brand.
- **Advocacy**—Consumers can conveniently communicate their liking of a new apparel or an entire collection to their friends and fans. Advocacy goes beyond merely preference about a purchase; it includes advocacy for the experience and brand overall and not just to close friends but a consumers' entire follower base.
- **Affiliation**—The membership benefits are enormous powerful and create loyalty to *Burberry*.
- **Identity**—Luxury fashion is about self-expression and identity building. The *Burberry* efforts deliver opportunities for self-expression because consumers can easily invite others in a viewing of a show. Consumers can select from the latest collection items and make recommendations to friend to express their own values, beliefs, and tastes.

The case study shows that by enabling the **six dimensions or social behaviors** that drive **social currency**, companies can build new business models and strengthen their value proposition to customers. What's more important is that we are just at the beginning of the enormous possibilities that become available through this new social age. To quote *Mary Meeker*, a Partner of a leading VC company: "The magnitude of the upcoming change will be stunning—we are still in Spring training."
Erich Joachimsthaler, (Vivaldi Partners Group)

A **guiding principle for establishing the social brand value** is represented by the statement that the best way to commit customers to a company or brand is to interconnect the customers. Such **branded communities** can increase the loyalty to the brand ideally and have a positive influence on the image. In the best case, these communities define themselves via the properties of the brand and internalize their values. Companies can contribute towards this by providing appropriate content. It is interesting to see that at this stage, increasingly larger parts of *Google* search results for the most famous brands refer to content created by users (cf. Searchmetrics, 2013).

The **process for building up brand behavior** to survive in the "social landscape" can obviously take place in line with the **SIIR model**, in order to initiate an appropriate **change process** (cf. Esch, Rutenberg, Strödter, & Vallaster, 2005, p. 995 f. for the basic principle). By means of the four phases sensibilize, involve,

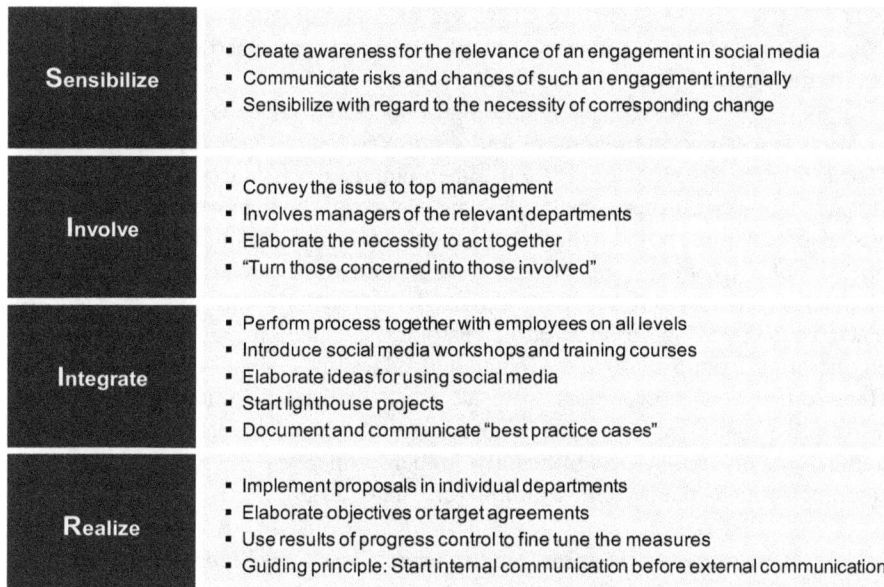

Fig. 9.11 SIIR model of a change management process (*Source* Author illustration)

	Michael Dell	Satya Nadella	Richard Branson	Steve Jobs
Facebook fans	52,000	110,000	850,000	7,400,000
Twitter followers	600,000	0	3,900,000	-

Fig. 9.12 Number of fans and followers of selected CEOs—approx. numbers (*Source* Author illustration, data source: Facebook/Twitter, 2014)

integrate, and realize, the aspired change process can be completed step by step in the company (cf. Fig. 9.11).

Here the question is asked—like with many other decisions having significant influence on the company—what position **top management** has towards an **engagement in social media**? The question can be given a simple answer. Are the top protagonists of the company present there—actively and with lifeblood? Examples of this may be presented by the CEOs of *Dell (Michael Dell)*, *Microsoft (Satya Nadella)*, and *Virgin (Richard Branson)*. The extent to which these appear to the outside, and to the inside as the case may be, can be seen by the number of the fans or followers. Figure 9.12 shows the image of the three CEOs named. Yet it is even more important to know to what extent they are spoken about. *Steve Jobs* has been incorporated as a benchmark. Such benchmarks are important in order to be able to make a comparison of one's own actions in a competitive environment. And it can also be checked whether CEOs are also participating via **hangouts**. This means video chats with a restricted number of participants. To this end, the users require the unique URL of the hangout.

Decisive for the sustainable anchoring of necessary changes is that the **criteria for assessing senior management and employees** are also appropriately adapted. If customer satisfaction or even customer enthusiasm is to be anchored in the company in daily business, then the compensation of the managers and employees must also be aligned to this criterion. Without such a consequent implementation—for example, by means of increased engagement in social media—customer orientation remains a mere lip service. In order to avoid this, at *Dell*, for selected managers, the proportion of customer satisfaction increases the variable remuneration from 5 to 40 % of the annual salary. Of course, this had significant effects on how working hours were allocated to various activities. If 40 % of the variable remuneration depends on customer satisfaction, then it goes without saying that, of course, a significantly higher proportion of time is invested in solving customer inquiries for example (cf. Buck, 2013).

> ▶ Remember Box **Senior management and employees must be rated on what the company expects of them**: Whether this is customer orientation, customer enthusiasm, service friendliness! Otherwise it won't work!

Think Box
- How seriously does my company take social media?
- How well are senior management informed about the chances and risks of social media—even in comparison to our competitors?
- Who of our management team is active on *Facebook* or *Twitter*?
- And how comfortable are they with these? Is the engagement in social media done from a sense of duty or with lifeblood?
- Is the CEO and/or the CMO or other managers responsible for marketing, sales, and communication active in social media—beyond a possible alibi presence?

> ▶ Food for Thought Until now, the saying applied—in particular for communicative tasks:
> **"The fish not the fisherman must like the bait."** Yet with regard to social media, it must now be imperatively supplemented: **"However, the fisherman must at least have fun fishing."** Otherwise the fish as in the customer will notice it—and swim away, irritated!

What could the **organizational implementation of social media marketing** look? Before a company engages in social media, it must be clarified where the

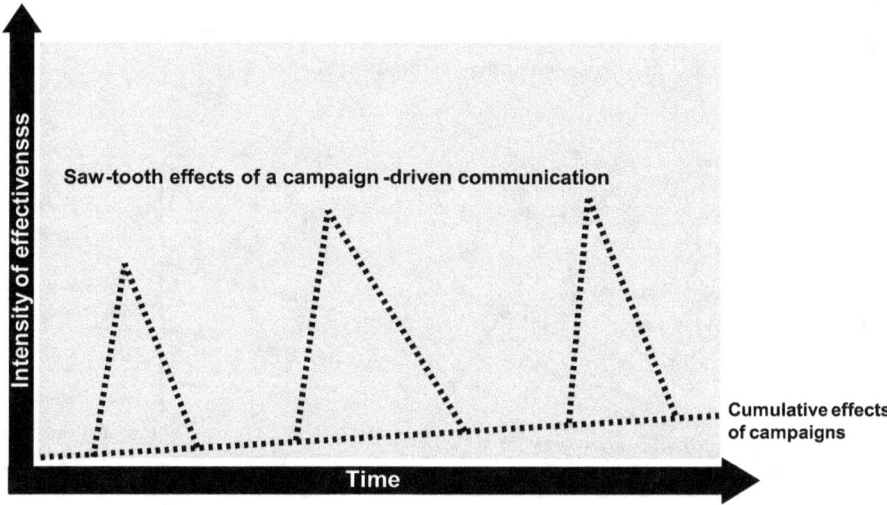

Fig. 9.13 Ideal type of effectiveness of a campaign-driven communication (*Source*: Author illustration)

responsibility involved needs to be anchored in the organization. In addition, human resources are required for social media marketing in the long term. When choosing, it must be borne in mind that the responsibility for social media and the "classic" dialog with customers should ideally lie in one hand. A requirement for the successful adoption of the corresponding responsibility is that the managers have recognized the **significance of the community** and its specific rules and are ready to confront them with appreciation. In the process, one needs to think with the head of this target group and feel with the heart of this target group to find the right "tone" for the interaction. Due to their affinity to social media, the managers to be defined are frequently representatives of the **digital natives**. Due to their own involvement, they have frequently recognized that marketing is particularly effective when companies are able to comprehend the social context. Especially digital natives think like marketers not in target groups or media plans. They try to understand and see which social networks a person is using. This encounter at eye level is an important requirement for acceptance for the engagement of a company in social media. However, it must be asked whether these digital natives have already established the necessary competences for the company to speak out in social media.

The **organizational structures** prevailing in companies—with regard to operational and organizational structures—must thus also change significantly because in the past, they were defined in a very campaign-driven manner. One campaign followed the next—and rarely had anything to do with the one before. The **serrated pattern of the campaign-driven communication** is illustrated in Fig. 9.13.

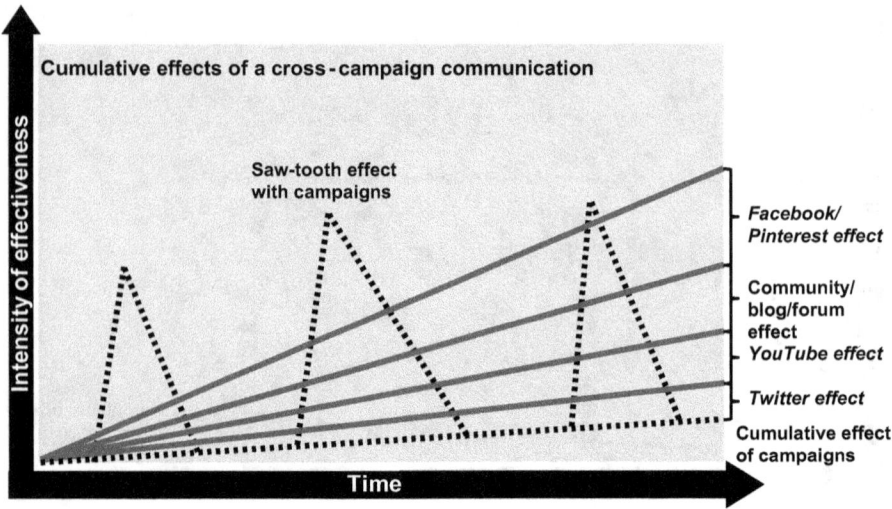

Fig. 9.14 Ideal type of effectiveness of cross-campaign communication by means of various engagements in social media—in comparison to campaign-driven communication (*Source* Author illustration)

This is increasingly being replaced by the necessity for **cross-campaign communication** which is indeed backed by continuous campaigns but reveals an increasing **communicative background noise**—even in times of no campaigns. Figure 9.14 shows the overall effect that can be achieved by this. The effects of such communication on the organization are dramatic because this type of engagement in social media does not fit in today's structures of many companies. In order to feed the various social channels, not only a long-term cross-channel and well-formulated **content strategy** as in an **editorial plan** is required but also the **personnel and financial resources** required for the implementation. A **conversation calendar** can be developed for the dialog in social media such as in the case of *Procter & Gamble* (cf. Einicke, 2012).

If such a dialog option is taken seriously, call center management must not run an **avoid contact strategy** according to which employees and management are assessed and paid as to in how short a time they have to "make a deal"—not to mention "break a deal"—re phone calls and other customer contacts. Furthermore, the question is posed here about the extent that the **management of customer contacts** should be outsourced. Against the background that, for example, *Facebook* users expect an answer to a post within 2–6 h, with *Twitter* inquiries it is about 2 h—whereby these response times are primarily expecting during regular business hours, this consideration is also relevant. The "expected" response time to emails, in contrast, is about 24 h.

> **Think Box**
> - What consequences are associated with saying farewell to stiff campaigns in my company?
> - What resistance is to be expected from what areas?
> - What solutions are there—on a process level as well as with operational organization?
> - Which effects can be achieved by cross-campaign communication?
> - How can the corresponding dovetailing be achieved?
> - Who can be the "driver" for this in my company?

The manager entrusted with social media marketing acts as an interface between the company and the users of social media. The **range of tasks the social media manager** has covers the following areas:

- Elaborating and communicating internally the **social media goals** as well as social media KPIs for them to be measured
- **Informing**—in particular the top management—**about the chances as well as the risks** of a social media engagement ("vaccinate the important top performers")
- Determining the need for resources (personnel, financially, structural) and procuring these **resources** in the company
- Elaborating a **social media concept**—in close interweaving with other marketing and communication measures of the company (incl. an editorial plan)
- Development of autonomous and integrated **campaigns** for use in social media
- Development, communication, and enforcement of the **internal and external social media guidelines**
- **Intra-communication** of the social media concept (incl. training courses and workshops)
- **External announcement of social media activities** and gaining of fans, followers, as well as users and/or members for own communities, forums, blogs, etc.
- Entering into **communication with the users** in blogs, forums, communities, and social networks—parallel to the **implementation of the editorial plan**
- **Analysis and possible seizing of proposals by the users** to take these to the company
- Ensuring of **web monitoring** (analysis of the own company, the own offers and brands or activities aimed at the own sector as well as appropriate monitoring of the competitors)
- **Reviewing the effects of the social media engagement based on informative KPIs**
- **Controlling the entire social media marketing** by means of corresponding budgets

- Ongoing ascertainment that the activities in social media are consistent with the **company and brand values**

An essential requirement for the social media managers to be able to accommodate this extensive range of tasks is a **"hotwire" to the specialist departments**. That is where the expertise on the company's services, which is indispensable for competent communication in social media, is anchored.

As social media marketing is still a relatively new approach, various concepts are under discussion regarding the **defining of responsibilities** and the **integration in the company structure**. Depending on the objectives of use in social media, the corresponding activities can be seen as part of the communication with customers or as public relations. There are various concepts when **integrating in the company organization.**

Decentralized anchoring of social media marketing
- In many companies, the responsibility for social media marketing is first and foremost the **result of organic growth**. Many organizational units take on subtasks with regard to the maintenance of social media. Besides the PR department, these can be the classic marketing unit, sales, and the customer service center. In order to find a consistent voice towards the stakeholders—in particular the potential customers and customers—extensive, cross-departmental coordination is required. The anchoring of direct customer contact—across the organization—opposes the risk of not being sufficiently quick-witted when increased customer service is required or in the event of a shitstorm, because the organizational requirements were not created for them.

Community manager
- A community manager monitors the **engagement on all social media platforms** and creates the content of it. This manager should be part of the marketing or sales department and thus be involved in areas close to the customer so that he is in close contact with the entire incoming and outgoing customer communications. His task is to communicate internally and externally and thus generate a **connection between the inside world of the company and the presence on the social map.**
- With that, the community manager is involved in an **area of conflict**. On the one hand, he should represent the company's interests as an employee to the outside. On the other hand, he must not act too marketing- or sales-oriented in order to be accepted by the community. There, it is more about giving the company a "human face" and to be touchable in a virtual sense without, however, denying one's own origin. For the community manager not only observes the various social media but also actively participates in them. To this end, he maintains own social media platforms and engages in relevant blogs and on relevant sites of the social networks.
- The bandwidth of the reactions of the company to be coordinated by the community manager ranges from the **active provision of information** to

counterstatements in the case of false reports which, for example, are discussed in communities or blogs for the first time. In addition, the advocates and opponents as in digital opinion leaders and influencers relevant for the company are to be identified in order to take particular care of them. The field of responsibilities also covers the **seizing of proposals for improvement for products and services** which are to be passed on to the corresponding departments within the company.

- If the responsibility of the community manager is upgraded and divided up among several shoulders in terms of personnel, a **hub and spoke concept** can emerge. "Hub" stands for the "crossroads" or "center" and "spoke" for the "crossing." This center can be formed by the community manager who supports and advises the other company units in handling social media tasks. By doing so, a holistic social media signature can be ensured even with decentralized responsibility.

Task force for social media marketing
- The anchoring of social media marketing can also take place in the form of a special task force in which those responsible from the departments entrusted with customer-oriented processes are involved. Their representatives have, besides their regular work, the task of shaping the presence of the company, its brands, and/or offers within social media and to gain relevant insights from them for the original field of responsibilities. To do so, the community manager's tasks described can be divided among several people. At the same time, it must be ensured that the **consistency of the external image** isn't lost, even if a certain **plurality of opinions in social media** can increase the credibility of companies.

Social media team
- Alternatively, a permanent social media team can be set up which focuses on the challenges of social media and is, for example, set up within the PR, marketing, or sales department or acts in close collaboration with those responsible for marketing and sales. In this case, the community manager's functions described are to be divided among several people.

Social media department
- A further variation for anchoring the social media engagement is represented by the setup of a corresponding department where the **social media decision-making competences** are centralized. This means that all activities associated with social media are coordinated via this unit. In doing so, it can be ensured that the company's image—across all social media—can be shaped holistically. In order for extensive networking to also be realized, this department is to be set up on a high hierarchic level and within close proximity to marketing or sales. This is the only way a consistency in customer service, in particular for potential customers and customers, but also other stakeholders, can be ensured,

irrespective of whether PR, sales, investor relations, or customer service staff is involved.

Holistic anchoring of social media marketing
- A consciously effectuated **decentralized anchoring of social media** marketing can be the **final step of the organizational implementation**. The responsibility for the dialog—especially with potential customers and customers—is no longer restricted to one or a few areas, but penetrates the entire organization. A requirement for this is, however, that all persons acting on the **social media front** are extensively trained in the corporate policy of the company as well as the use of social media. Apart from that, there must be meaningful internal social media guidelines which are observed, in order to ensure a consistent overall image of the company in spite of this decentralized responsibility.

An **external initial consulting** or with larger companies, the **setup of an internal social media consulting** could be constructive alongside the organizational anchoring, in order to inform the entire organization and especially the employees working at the customer interfaces extensively about the operative factors and fields of application of social media and to constantly train them. In addition, internal training courses and workshops are to be conducted to sensibilize and win the entire organization to the issue.

Irrespective of which **type of organization** is chosen, every company must—in parallel to the internal training courses mentioned—elaborate **social media guidelines** before engaging in social media and communicate them internally. They are to govern how employees of the company are to behave with regard to their engagement in social media. The guidelines should make it clear what every employee may do and say and what restrictions apply in the scope of externally directed communication. The company should also think about how **private engagement** and **engagement for the company** can be combined for the employees. With the help of internal social media guidelines, the company can indeed influence what the employees circulate, for example, in blogs (like *Tumblr*) or on *Twitter* or *Facebook* on behalf of the company but not what issues they communicate and how as a private individual. Here it is becoming increasingly apparent that professional and private matters are intermingling more and more. For this reason, it has become necessary to "guide" one's own employers in terms of engagement in social media in order to protect the company, its brands, products, and services in this way. Social media guidelines make a significant contribution towards creating awareness for engaging in social media.

In the **internal social media guidelines** for the own employees, the behavior guidelines for the following areas are constructive:

Definition and communication of social media objectives
- The objectives of corporate engagement in social media are to be elaborated and made transparent for all employees. The bandwidth of these objectives ranges

from the distribution of general company news over the promotion of selected products/services (e.g., in social networks or via engagement on *Twitter*) to the setup of corporate blogs as well as online forums and communities. It must also be clarified which target groups are to be focused on. These can be consumers, suppliers, cooperation partners, and/or opinion leaders as well as the own employees.

Safeguarding the necessary confidentiality of internal matters
- The distribution of company and business secrets, from information about ongoing projects (e.g., technological developments) over the status of ongoing acquisitions, of financial data as well as information about business partners, customers, and employees simply must not happen. This means nothing other than that the responsibilities and duties of the employees continue to hold. Deviations from this require previous written permission by the executive board. With that, the guidelines possibly already anchored in the company re offline communication are extended into online media. This includes, for example, official inquiries by representatives of the media or issues of a legally relevant nature by the responsible PR or legal department. The PR department is still responsible for official company announcements.
- In their code of behavior, *Hewlett-Packard,* for example, expressly forbid their employees to make comments about the following issues: business results, intellectual property, court proceedings, changes in management, dismissals, matters concerning shareholders, as well as contracts with business partners, customers, and suppliers. *Hewlett-Packard* not only relies on prohibitions but also calls upon their employees' sense of responsibility.

Authenticity of the persons acting on behalf of the company
- To be accepted in social media, it is important to achieve great **credibility** as a communicator. For this reason, employees acting on behalf of the company make the origin obvious by stating their own name, their position, and the company. By doing so, they become recognizable and approachable as a **representative of a company**. These data can be communicated via legal info or the profile description. If employees point out in their own posts whether they are presenting their own or the company's opinion also contributes towards credibility. This can reduce situations of conflict. For this reason, so-called fake accounts should not be set up. These include user accounts on *Facebook* or *Twitter,* for example, where the user hides his real identity to act from an anonymous position and advance certain views.

The person communicating is responsible
- As it is with offline communication, it also applies for online communication that each communicator is responsible himself for the effects of his actions. The employees must thus be given ample information about what effects views advanced online can have in comparison to offline statements. Views advanced online not only spread much faster but are quasi-viewable by everybody and can

hardly be removed from the Net. For this reason, the possible consequences of false and/or reputation-damaging statements (e.g., about competitors, customers, suppliers, colleagues) are unpredictable.
- This rule not only applies to posts by employees on behalf of the company but also to posts made as a private individual. It especially applies to such content giving the impression that a private individual is speaking on behalf of the company. For this reason it is imperative, unless otherwise stipulated, to communicate messages in the first person. At the same time, it must be taken into consideration that there is no longer protected privacy on the Internet. Employees must always be aware that they are quasi never **merely** a private individual on the Internet but that connections can also always be ascertained to professional engagements. The **duty of loyalty as per labor law** also continues to have effect in private life. That is why it must be made clear that social networks and the engagement of employees in them do **not** automatically count as privacy. Consequently, only such content that can be signed by one's own name with a good conscience should be made public in all social media.

Professionalism in actions
- A professional appearance includes clearly differentiating between the presentation of facts and opinions ("I am or company XY is of the opinion that...") in comments. In this way the **credibility** of statements is greatly increased. A part of professionalism is also **fair play** that goes without criticizing competitors. This may include—like in the case of blog entries—an **internal dual-review principle** prior to publishing by means of a critical look, in order to avoid posts evolving from high emotionality being published that may be later regretted.

Internal criticism remains internal
- At any rate, it must be avoided that employees circulate their "frustration" about their own company, customers, suppliers, or colleagues via social media. Towards the outside, it is essential that every employee should stand behind their company, its brands, and offers. If this is not the case, to keep silent can be the appropriate **form of solidarity** of the employee with the company. As tempting as it may be, internal criticism should not be circulated under a pseudonym as such cover-up tactics by Internet users are frequently uncovered and can lead to negative consequences for those concerned.

Dealing openly with mistakes in online posts
- If wrong or unsuitable online posts of own employees are identified, then these should be corrected actively. Corresponding entries in blogs, forums, or communities, however, should not be altered or removed without comment. In this case it is essential to enter into an **open dialog** between the employee and the superior in order to make clear the risks of the entries as well as the necessity to correct them. To this end, appropriate **escalation mechanisms** are to be

developed. It may even be necessary to justify the necessary corrections towards the online public. This can lead to the authenticity of actions necessary.

Defining responsibilities for social media
- In order to prevent a communicative "uncontrolled growth" in a company caused by an uncoordinated communication of many internal senders, the responsibilities for the social media engagement need to be clarified in the company. For this purpose, it must first be determined which employee(s) is responsible for posts via *Twitter*, on *Facebook*, in external blogs or in the corporate blog, etc. This responsibility can also be allocated according to expertise. On the other hand, it must be defined who may operate, for example, an own blog on company issues. This decision should not be left up to the employees themselves. For this reason it must also be defined from whom **"permission" for engaging in social media** on the company's or brands' behalf can be obtained. In addition, it must be defined under which conditions such permission can be denied. Likewise it must be defined who, for example, is responsible for **seeding** with the allocation of certain topics. Even if it is difficult to define general requirements, **guidelines about the form of integration and participation of employees** with different topics should be created.

Using social media during working hours
- Under heavy discussion is the question as to whether companies should stipulate to what extent an engagement in social media should be allowed during working hours. However, if a company wants to be "skillfully" active in the social landscape, then it can seem strange to forbid employees responsible handling during working hours. But here it also applies: Trust in employees must also be justified!

Observing applicable laws
- Every employee who is active in social media must be made fully aware of the regulatory framework surrounding his actions. It must be particularly pointed out that the laws governing data protection, copyrights, brand rights, and personal rights do not lose their validity in the online environment—quite the contrary. These aspects are thus given special attention because:

> ▶ The Internet forgets nothing!

In the course of the organizational anchoring of social media marketing, the **definition of the responsibility for such social media guidelines** also takes place. These cover the initial development of these guidelines, the continuous amendment of them, as well as the internal publication. This also includes the task of committing the employees to the social media guidelines. The responsibility for this can be located in social media organization and/or in human resources. In addition, sanction modalities are to be defined which then take effect when employees do not observe these social media guidelines. The shop council may be involved.

Basically, the aim of social media guidelines is to achieve a **controlled own initiative of the employees**. As it became obvious, this is only apparently a contradiction.

If a company is the **initiator of social media platform** itself (e.g., of a blog, an online forum, or an online community), it makes sense in advance of a corresponding engagement to draw up and communicate *external* **social media guidelines** for the external users of these company media. Answers to the following questions should be included in them:

- Is the community moderated by a company representative?
- May the Internet users say anything they want to in blogs, communities, forums, and other company platforms?
- Can impolite posts and those drifting off from the subject be removed?
- How should insupportable comments by users be handled?
- How can false reports be rectified?
- Are posts by users reviewed by own staff before being published and are unsuitable posts excluded from being published (e.g., abusive criticism, putting down competitors, etc.)?
- May anonymous comments be posted or is a log-in necessary?

These **guidelines** serve as **etiquette of social media**, to be referred to in the case of a crisis. They can at least reduce tensions between users during interactions. At the same time it is an act of fairness to communicate them in advance.

Think Box
- Where is my company positioned with the implementation of a social media strategy?
- Have the tasks for those responsible for social media been clearly defined?
- Has an intensive coordination with the specialist departments been ensured?
- What form of integration of social media in the company is given (decentralized anchoring, community manager, task force, social media team, social media departments, holistic anchoring)?
- Have internal social media guidelines been drawn up and communicated extensively?
- Are there also corresponding social media guidelines for our external social media activities?
- Who bears overall responsibility for this?

Yet the question is: How well are companies prepared for **social media marketing** and the necessary **digital transformation**? Companies frequently have no fixed guidelines when it comes to the response time to inquiries or comments by the users. Long waiting times can result in customer dissatisfaction. And this also applies outside "normal hours of business." In line with the statement "**social media never sleep,**" the responsibility for monitoring social media cannot simply be stopped over the weekend because that is exactly when a shitstorm wave can develop. After being inactive in social media for 64 h, the company *Vodafone*, for example, was surprised on Monday morning and was already standing in the middle of a shitstorm that was able to develop over the weekend without being noticed, stopped, or commented on by *Vodafone*. Even here, *Vodafone* learned and spotted "silence" as being a mistake. This is the reason why social media presences are now monitored around the clock. Suspicious movements on the website are identified by an **early warning system**; responsibilities and decision-making processes are clarified in crisis situations.

One essential requirement to protect oneself against attacks in social media or to respond to them in an appropriate manner is that companies become aware of waves of negative propaganda as early as possible. Thus by means of the previously addressed **web monitoring** or also **online trend monitoring**, the relevant forums, blogs, communities, etc., should be examined for buzzwords such as rip-off, boycott, PR, lie, "sucks," etc., in association with the own company and/or brand name. If negative rumors appear on the Internet or if products and services are rated extremely badly on rating platforms, companies should be informed about this at an early stage in order to be able to react correspondingly (cf. Chap. 3).

> ▶ **Remember Box** Social media are a powerful reinforcer of word-of-mouth advertising, and a turbo for negative PR at that.

How did, for example, *WWF* position themselves after the shitstorm they survived (cf. Chap. 6)? A new communications department with a specific social media editor was established. Furthermore, duty and operation schedules were drawn up for emergencies. *Nestlé* has installed a **Digital Acceleration Center** to monitor the buzz in social media and to react accordingly and fast (cf. Blackshaw, 2013).

> ▶ **Food for Thought** This insight is important: **A shitstorm cannot be prevented if you are not present in social media.** Being present there may alleviate the start of a shitstorm. However, as a company, you then do also have a "practiced" channel to stand up to the attacks. This is where the—slightly altered—saying holds true once again:

Anybody who buries their head in social media "sand" needn't be surprised if they grit their (digital) teeth.

In order to appropriately capture the effects of social media, a **social media balanced scorecard**, as can be found in Fig. 9.15, is needed (also cf. Fiege, 2012).

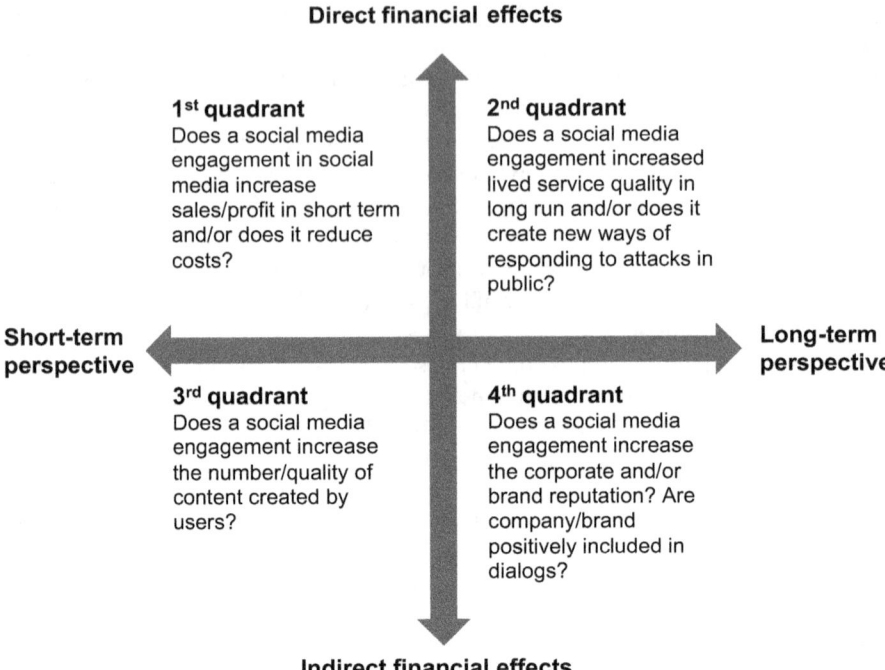

Fig. 9.15 Areas of questions in a social media balanced scorecard (*Source* Author illustration, data source Ray, 2010)

With that, two axes must first be differentiated. The y-axis shows the direct and indirect **financial implications of a social media engagement**. The x-axis illustrates the **time perspective** and demonstrates whether the corresponding effects will emerge rather in the short term or long term. A differentiated analysis is the only way to clearly capture the various effects. After all, an activity in social media often directly leads to financial results—such as more sales, higher profits, or reduced marketing costs (**1st quadrant**). It takes a long time for some effects to impact monetary indicators. This can be expected with, for example, improved services and/or a positive awareness of companies and brands in public (**2nd quadrant**). The various types of content generated by users can be found in the **3rd quadrant**. Frequently, these only have an indirect impact on a company's financial control indicators if, for example, impulses for product development are given. These effects can appear in the short term, but in part also delayed. Finally, in the **4th quadrant** are those effects that principally have a positive impact on the reputation of company and/or brands in the long run.

By means of a **balanced scorecard** it needs to be reviewed in terms of the company, for example, how important the number of *Facebook* fans, the number of positive retweets, the number of visitors to the own social media representation, the number of video views, (positive) ratings, (positive) entries in blogs, and communities are. The fact that some effects do not have an impact until in the long run must not lead to these potential-oriented effects being neglected merely because they are not directly visible in the profit and loss statement or in the balance sheet.

The additional communication channels and tools first and foremost require two things across the company: **coordination and integration**. It becomes hazardous for companies when further departments and agencies, which are responsible for app marketing, social media marketing, SEO, SEA, and online advertising, are added to the already acting departments and agencies for traditional advertising, dialog marketing, sponsoring, internal communication, investor relations, event marketing, corporate identity, exhibition engagements, and corporate publishing. The risk increases when both the own departments and the agencies act more or less independently of each other, and each tries to develop and sell—from their point of view—the "ideal" campaign idea.

The goal to be strived for is called **consistency** across all channels, tools, and agencies in order to achieve a coherent overall image of the company. All internal and external measures—whether online or offline oriented—must be geared to the main goals of the company in order to generate a **coherent company, brand, and/or offer identity**. In order to achieve this consistency in the case of cross-media campaigns via online and offline media, it is advisable to pool the agencies right from the start. An approach can be more constructive, where all agencies responsible for communication receive a **briefing** at the same time and place, the basis of which is then subsequently worked on together. This briefing is to be elaborated in a **holistic approach** across all departments. The **safeguarding of consistency** thus takes place in the process of creating the briefing and the development of the campaign based on it. After that, the results are presented within **one presentation document**. With that, it is essential to overcome the petty jealousies with regard to one's own responsibilities—internally and externally in equal measure. "And one wishes the agency scene that not every specialist believes to have solved all communication problems the company or brand has with his individual solution" (Mayer-Johanssen, 2012, p. 24).

This type of "forcing consistency" can be accompanied by intracompany **campaign management** which—by means of integrated concepts—in turn forces the achievement of integration beyond individual communication channels and tools, where the customer himself and not individual elements or channels of a campaign are the main focus. This is the only way a convincing customer journey becomes possible, even if it does lead past many customer touch points.

> ▶ **Food for Thought** The challenge is to set up a **holistic KPI-driven real-time marketing and management process** that reveals high flexibility within the defined strategic corridor of development for the company. In doing so, it represents an indispensible task of marketing to seize the leadership role offered and to substantiate sustainable value creation in the company by contributions.

The challenges there with regard to **reconsidering existing management tools** are demonstrated in the short article *Value Constellation*.

> **Article by Guest Author *Clark N. Banach***
>
> **Value Constellation: Making Models Work for You!**
>
> We have reached a point where we have been able to develop a **parallel marketplace beyond the limitations of the physical world**. This marketplace has evolved to have its own supply chain and commercial environment. Although the digital and physical landscapes have developed into independent economies, they are far from mutually exclusive. The challenge occurs when the **digital landscape** evolves at a rate that dramatically exceeds the rate of the evolution in the physical world. This instability has led to majestic and often unprecedented interaction. In order to react to the rapidly evolving landscape, organizations must adapt traditional models to new uncharted scenarios. The models must be adjusted to meet the complexities of the landscape. It is essential for businesses to develop adaptations of traditional models in order to stay competitive. It is also required to navigate the complex and evolving environment created by the interaction of the digital and physical landscapes. Managers must take it upon themselves to create process models that can take advantage of rapidly emerging opportunities.
>
> For example, one could apply *Ansoff's* **Product and Market Development Matrix** (cf. Ansoff, 1957, pp. 113–124) to the context of digital products and marketplaces of the artificial. The model, which traditionally produced 4 strategic orientations, now can produce 16 (cf. Fig. 9.16). The modification provides a more specific **scenario analysis** that can act as a clearer lens to evaluate preexisting portfolios and future growth opportunities.
>
> The **different strategic orientations** require comprehensive analysis and benefit from unique and detailed approaches to the marketing mix. For instance, a **new digital product** being developed for a **new digital market** would have different challenges than a **new tangible product** being brought to a **new digital marketplace** (cf. Fig. 9.17). The most dramatic differences exist in **supply chain management** decisions; however approaches to promotion decisions could be compromised if not made in the clearest possible context. The *Creative Suite* and *Master Collection* Software produced by *Adobe Systems* is an example of how an existing digital product sold in an existing tangible marketplace was successfully migrated to a new digital market owned and operated by *Adobe Systems*.
>
> In contrast, *Apple* developed a strategy to place *iTunes* downloads, an existing digital product, in new tangible markets by placing *iTunes* gift cards in checkout lines of dominant global retailers. Strategic planners and project managers will reduce the risk of uncertainty by striving to become **champions of innovation**. Planners cannot wait for the emergence of industry-accepted models to reach their desks before they begin to adjust to the rapidly evolving context in which they are making decisions. When an organization can modify traditional models or create new agile models, it develops **competitive intelligence** that can be

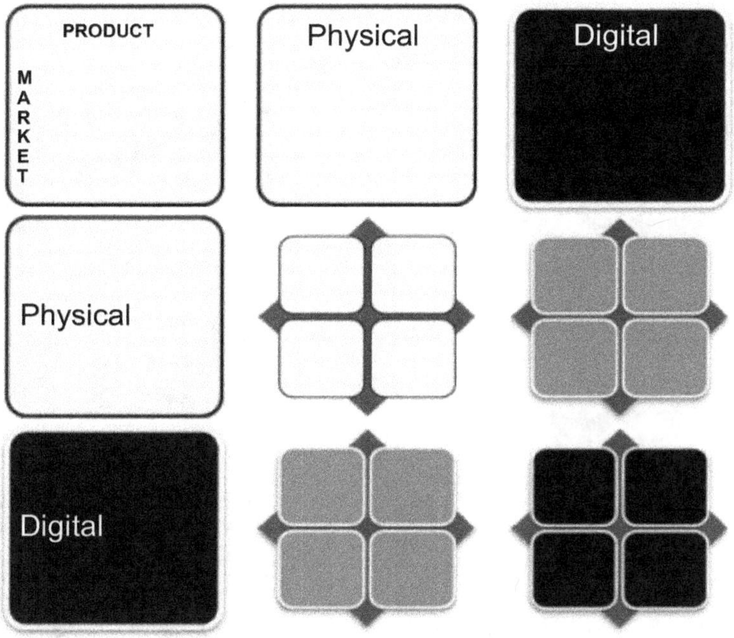

Fig. 9.16 Digital product matrix (*Source* Illustration by Banach)

leveraged to generate **sustainable competitive advantages**. The collective intelligence of the organization can then be systematically organized, prioritized, and distributed to those involved in **value creation** in more effective ways.

Making Value Constellations

The **value constellation** is a modification to the **value chain** of *Porter*. It shapes an approach to project management, operations management, and supply chain management for the rapidly shifting and innovative service industry. The model offers insight to analysis and practical approaches for a wide range of project types. This includes standardized projects (as is the case with web development and design) and projects with unique and uncommon structures (as is the case for incubating a trust to save a local park). The **value chain** allows us to identify points of parity and core competencies in our organization and supply chain (cf. Fig. 9.18; Porter, 1996, pp. 61–78).

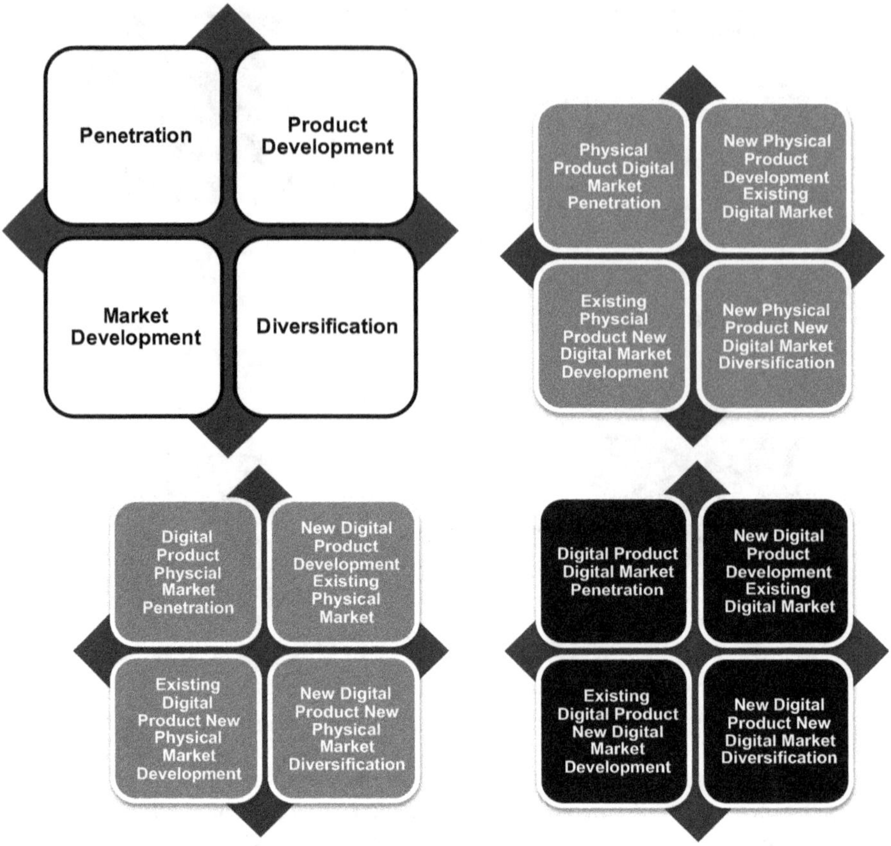

Fig. 9.17 Digital/tangible market/product development matrix (*Source* Illustration by Banach)

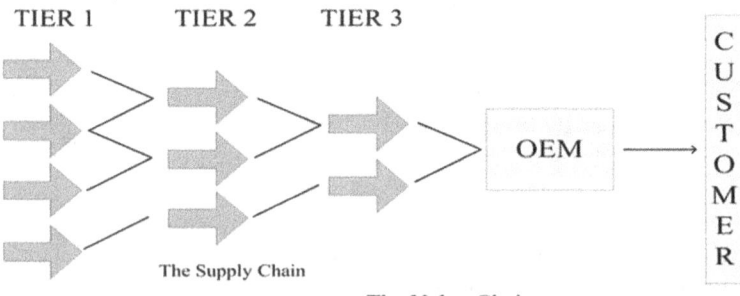

Fig. 9.18 Supply chain consisting of value chains (*Source* Illustration by Banach, data source Waissi, 2011)

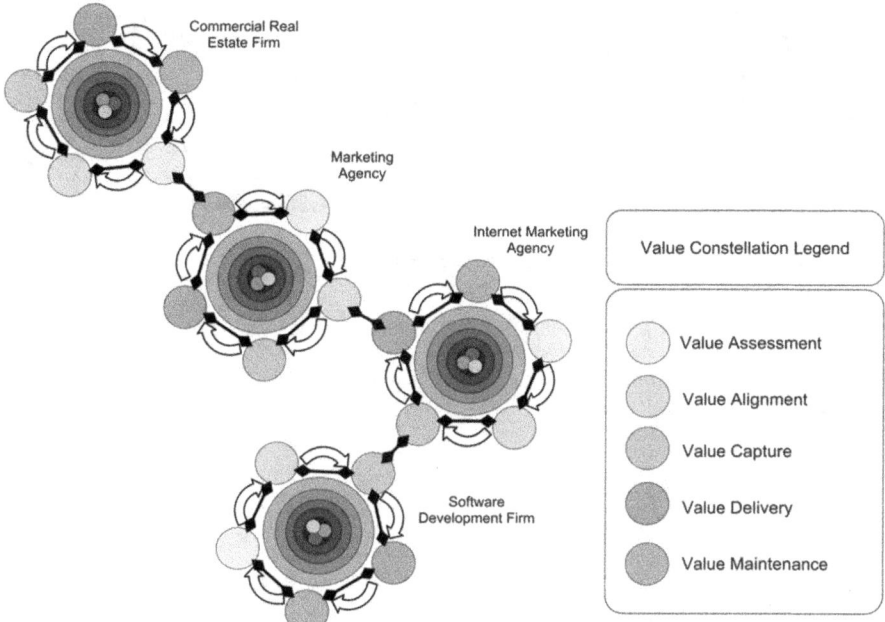

Fig. 9.19 Value constellation (*Source* Illustration by Banach)

The **value constellation** aims to do the same for businesses that do not follow linear paths to value creation in knowledge-based economies (cf. Fig. 9.19). The value constellation creates a foundation for projects of all types whether they are simple with short timelines or complex with ongoing development. Whether it was dental services, legal consulting, software development, or promotion engagements, the value constellation was capable of handling work packages and providing insight in each new environment. The value constellation can be complemented by appropriate management practices. When a communication strategy is in place, supported by appropriate infrastructure, the framework becomes capable of **handling the complexity of a nonlinear value chain** in a knowledge-based economy.

The **original value chain model** requires an underlying assumption that industries are homogeneous in nature and that organizations operate along broadly similar lines in a context where industries are relatively stable in nature (cf. Fig. 9.20; cf. Helm & Jones, 2010, pp. 579–589). The value chain model works well for organizations that convert raw materials into finished products, in order to produce goods through standardized activities, but many successful modern businesses do not manufacture or sell goods.

Despite current variations developed for service-based and network organizations, the value chain is still insufficient in offering guidance to **knowledge-based organizations**. In these cases where knowledge is a primary

Fig. 9.20 Recent adaptation of value chain (*Source* Illustration by Banach, data source Homburg et al., 2009, p. 59)

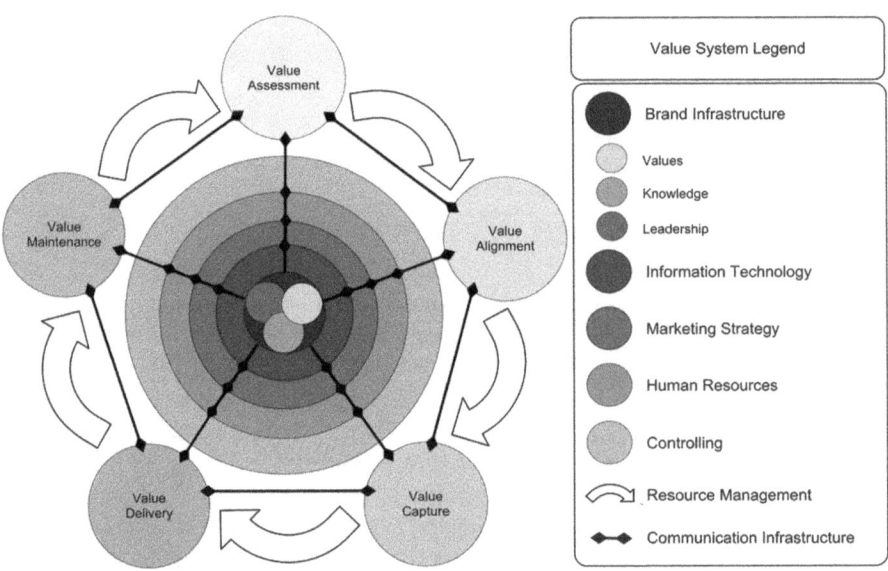

Fig. 9.21 Value system (*Source* Illustration by Banach)

resource and inquiry is the raw material, new solutions need to be explored in order to align the resources of an organization to solve complex problems (cf. Helm & Jones, 2010, pp. 579–589; Woiceshyn & Falkenberg, 2008, pp. 85–99). The solution is the creation of a **value system** (cf. Fig. 9.21).

The **value constellation** is meant to serve the same purpose as the value chain. Secondary functions are used to create the value required for demand generation. Primary functions are used to control resources that manage that demand, capture value, and ultimately fulfill demand (cf. Homburg, Kuester, & Krohmer, 2009, p. 59). In order to provide a cross-functional model, the

SCOR	Value Chain	Value System
Plan	Sales & Distribution	Assessment
Source	Inbound Logistics	Alignment
Make	Production	Capture
Deliver	Outbound Logistics	Delivery
Return	Service	Maintenance

Fig. 9.22 SCOR, value chain and information supply chain comparison (*Source* Illustration by Banach, data source: Baltacioglu, Ada, Kaplan, Yurt, & Kaplan, 2007; Homburg/Kuester/Krohmer, 2009, p. 59)

elements of the Supply Chain Council's SCOR (plan, source, make, deliver, and return) were implemented to increase odds of standardization and scalability. Figure 9.22 is a table illuminating how the preexisting models have been blended together to create a new approach. In organizations that participate in the delivery of services, a sale is often required prior to the execution of the service that will eventually be delivered. By moving the needs assessment to the beginning of the process, it accounts for the complex interactions prior to the sale in **knowledge-based service industries**.

Making Models Relevant
At the end of the day, if an organization is going to create, communicate, and deliver value in a rapidly evolving landscape, consisting of a new digital dimension, it must be prepared to develop new lenses to see the world. The **competitive landscape** is changing so dramatically that businesses cannot depend on the strategic models of the past or wait for new models to be accepted by the industry. The models provided in this example are meant to illustrate the opportunities that exist in the **modification of traditional ideas** to meet the **needs of modern paradigms**. If a business is going to remain agile and responsive to the challenges of these paradigms, innovative strategic planning is compulsory for survival. This article is meant to illustrate that it is not the answers you have, it is the questions you ask that will keep you in the game. It is the manager's obligation to develop models of his own and strive for excellence on behalf of those that trust his guidance.

C. N. *Banach*, (MBA, MA, Interactive Marketing Professional)

Quick Wins

- _____
- _____
- _____
- _____
- _____
- _____
- _____
- _____
- _____
- _____
- _____
- _____
- _____
- _____
- _____
- _____
- _____
- _____
- _____
- _____
- _____
- _____
- _____
- _____
- _____
- _____
- _____
- _____
- _____
- _____

Appendix

> The average gives the world its existence,
> the extraordinary its value.
> Oscar Wilde

References

Accenture. (2012). *Mobile Web watch*, accenture.com/SiteCollectionDocuments/PDF/Accenture-Study-Mobile-Web-Watch-Germany-Austria-Switzerland-EN.pdf, 8.2014.
Andersen, C. (2009). *The long tail*. München: Der lange Schwanz.
Ansoff, I. (1957). Strategies for diversification. *Harvard Business Review, 35*(5), 113–124.
ARD/ZDF. (2012). *ARD/ZDF-Onlinestudie 2012*, Mainz/Frankfurt.
Ascend2. (2013). *Marketing strategy report: Social media*, ascend2.com/home/wp-content/uploads/Social-Marketing-Strategy-Report-Ascend2-Edition.pdf, 28.2013.
Ashton, K. (2009). The 'internet of things' thing. *RFID Journal*, 22.7.2009. www.rfidjournal.com/article/print/4986, 2.10.2012.
Awareness. (2012). *The state of social media marketing report. 7 major findings and in-depth analysis*, September 2012, docs.google.com/file/d/0BzFS4FSdl83hdjJ0T2NVQmZvemc/edit?pli=1, 20.5.2013.
Baltacioglu, T., Ada, E., Kaplan, M., Yurt, O., & Kaplan, Y. (2007). A new framework for service supply chains. *The Services Industries Journal, 27*(2), 105–124.
Barnes, N. G., Lescault, E. M., & Andonian, J. (2012). *Social media surge by the 2012 Fortune 500: Increase use of blogs, Facebook, Twitter and more, college of business center for marketing research (CMR)*. University of Massachusetts Dartmouth, umassd.edu/cmr/socialmedia/2012fortune500, 20.5.2013.
Bauder, D. (2012). *Study shows growth in second screen users*, tv.yahoo.com/news/study-shows-growth-second-screen-users-122518350.html, 26.5.2013.
Baudis, M. (2012). *Best practice: FCB alert – Fan-Artikel nach Maß*, smartservice-blog.com/2012/05/24/best-practice-fcb-alert-fan-artikel-nach-mas, 24.5.2012.
Bauer, W. (2012). zitiert in Brauck, M. (2012). Fürchtet euch nicht!. *Der Spiegel, 45*, 146–148.
Baumann, A. (2012). Mathias Döpfner: Qualität setzt sich durch. *Bonner Generalanzeiger*, p. 5, 26.10.2012.
Berger, C., Blauth, R., Boger, D., Bolster, C., Burchill, G., DuMouchel, W., et al. (1993). Kano's methods of understanding customer-defined quality. *Center for Quality of Management Journal, 2*(4), 3–36.
BITKOM. (2012). *Social Media in deutschen Unternehmen*, Berlin.
Blackshaw, P. (2013). How digital acceleration teams are influencing Nestlé's 2,000 brands. sloanreview.mit.edu/article/testing-123, 8.2.2014.
Blumer, H. (1969). *Symbolic interactionism: Perspectives and method*. Englewood Cliffs, NJ: Prentice Hall.
Buck, M. (2013). *Verstehen, was Kunden bewegt: Der Kunde wird zum Co-Designer, Meinungsbilder und Markenbotschafter – Die Umsetzung des Empfehlungsmarketings bei Dell*. Vortrag auf dem Dialogmarketing Gipfel 2013, Frankfurt, 21.8.2013.
Bullas, J. (2012). *10 insights into the state of social media in 2012*, jeffbullas.com/2012/12/07/10-insights-into-the-state-of-social-media-in-2012/#comments, 16.5.2012.
Byron, E. (2011). In-store sales begin at home. *The Wall Street Journal*, 25.4.2011.
Campillo-Lundbeck, S. (2014). Allmächtiger algorithmus. *Horizont, 5*, 17.

Caputo, A. C. (2011). *Visual story telling*, visualstorytelling.com/chapter1.html, 14.11.2012.
China Internet Marketing. (2014). *China's top social media apps in 2013*, chinainternetwatch.com/5964/chinas-top-social-media-apps-in-2013/#more-5964, 25.1.2014.
Chui, M., Löffler, M., & Roberts, R. (2010). The internet of things. *McKinsey Quarterly*, mckinseyquarterly.com/article_print.aspx?L2=4&L3=116&ar=2538
Cisco. (2013). *Cisco visual networking index: Global mobile data traffic forecast update, 2012–2017*, cisco.com/en/US/solutions/collateral/ns341/ns525/ns537/ns705/ns827/white_paper_c11-520862.pdf, 8.4.2013.
Claus, A. (2014). Eine Waage für die digitale Diät. *Bonner Generalanzeiger*, p. 7, 8./9.2.2014.
Cluetrain. (1999). *Cluetrain Manifesto*, cluetrain.com/cluetrain.pdf, 2.10.2012.
ComScore. (2011). *It's a social world, Top 10 need-to-knows about social networking and where it's headed*, wearesocial.net/blog/2011/12/comscore-2011-social-world, published by Plasseraud, S. (2011). As a blog post under the title *"comScore 2011: It's a social world"*.
ComScore. (2012). *State of the Internet 1st Quarter 2012*, de.slideshare.net/indigitalmd/comscore-state-of-us-Internet-q1-2012, 25.5.2013.
Consultants, C. M. (2012). *Die Veränderungsdynamik des digitalen Marketings*. Studienergebnisse, Mannheim: Die vertagte Revolution.
D'Inka, W. (2012). Die Rundschau. *Frankfurter Allgemeine Zeitung*, p. 1, 14.11.2012.
DGUV. (2010). *Beim Multitasking sind alle gleich – schlecht, Studie des Instituts für Arbeit und Gesundheit der Deutschen Gesetzlichen Unfallversicherung (IAG)*, Berlin
Disselhoff, F. (2011). *Die peinlichsten Facebook-Pannen*, meedia.de/Internet/die-peinlichsten-facebook-pannen/2011/06/06.html, 6.6.2011.
Drucker, P. (1957). *Landmarks of tomorrow: A report on the new post-modern world*. New York: Harper & Row.
Duggan, M., & Smith, A. (2013). *Social media update 2013, Report Social Networking*, pewinternet.org/Reports/2013/Social-Media-Update.aspx, 4.2.2014.
Ebizma. (2013). *Top 5 most popular comparison shopping websites – April 2013*, ebizmba.com/articles/shopping-websites, 23.6.2013.
Eckerson, W. (2012). *The Two Sides of Facebook Intelligence*, b-eye-network.com/blogs/eckerson/archives/2012/03/the_two_sides_o.php, 16.11.12.
Einicke, B. (2012). *Dialog im Social Web – Das Kundenbindungsprogramm „for me" von P & G*. Vortrag auf dem Dialogmarketing-Gipfel, Frankfurt.
eMarketer. (2011). *Two in five mobile owners use internet on the go*, emarketer.com/Article/Two-Five-Mobile-Owners-Use-Internet-on-Go/1008553, 8.4.2013.
eMarketer. (2012). *Where in the world are the hottest social networking countries?, With 1.2 billion users worldwide, social networking usage patterns vary by country and region*, www.emarketer.com/Article/Where-World-Hottest-Social-Networking-Countries/1008870, 16.5.2013.
Esch, F.-R. (2005). *Moderne Markenführung, Grundlagen – Innovative Ansätze – Praktische Umsetzungen* (4th ed.). Wiesbaden: Gabler.
Esch, F.-R. (2012). *Customer touchpoint management*. Saarlouis: Berührung mit dem Kunden.
Esch, F.-R., Rutenberg, J., Strödter, K., & Vallaster, C. (2005). Verankerung der Markenidentität durch Behavioral Branding. In F.-R. Esch (Ed.), *Moderne Markenführung* (pp. 985–1008). Wiesbaden: Gabler.
Etsy. (2014). *Etsy at a glance*, www.etsy.com/press, 24.1.2014.
Ewens, S. (1996). *PR – A social history of spin*. New York: Basic Books.
Facebook. (2012a). *Info*, www.facebook.com/facebook?v=info, 15.10.2012.
Facebook. (2012b). *Facebook platform policies*, developers.facebook.com/policy/, 27.11.2012.
Fiege, R. (2012). *Social media balanced Sorecard, Erfolgreiche Social Media-Strategien in der*. Wiesbaden: Praxis.
Forrester. (2013). *Few consumers trust social media marketing, internet ads*, marketingcharts.com/wp/interactive/few-consumers-trust-social-media-marketing-internet-ads-28061, 31.1.2014.
Forster, F. (2012). *Social Media – vom Hype zum festen Bestandteil im Kundendialog*. Vortrag auf dem Dialogmarketing-Gipfel, Frankfurt, 22.8.2012.

Gallagher, L., & Zoratti, S. (2012). *Precision marketing: Maximizing revenue through relevance* (pp. 167–174). London: Kogan Page.
Gallup. (2012). *Engagement Index Deutschland 2011*, Berlin.
Gallup. (2013). *State of the global workplace, employee engagement insights for business leaders worldwide*, gallup.com/strategicconsulting/164735/state-global-workplace.aspx, 8.2.2014.
Gantz, J., & Reinsel, D. (2011). *IDC IVIEW, extracting value from chaos*, www.emc.com/collateral/analyst-reports/idc-extracting-value-from-chaos-ar.pdf, 23.1.2014.
Gartner. (2012). *Gartner's 2012 hype cycle for emerging technologies identifies „tipping point" technologies that will unlock long-awaited technology scenarios*, gartner.com/it/page.jsp?id=2124315, 12.9.2012.
Gartner. (2013a). *Hype cycles*, gartner.com/search/site/freecontent/simple?typeaheadTermType=&typeaheadTermId=&keywords=hype+cycle+emerging+technology, 20.1.2014.
Gartner. (2013b). *Gartner says mobile app store will see annual downloads reach 102 billion in 2013*, gartner.com/newsroom/id/2592315, 8.2.2014.
Gartner. (2013c). *Gartner says the internet of things installed base will grow to 26 billion units by 2020*, gartner.com/newsroom/id/2636073, 8.2.2014.
Geisler, B. (2012). *Adidas Neo in Hamburg: Alle mal hergucken*, abendblatt.de/wirtschaft/artikel2174532/Adidas-Neo-in-Hamburg-Alle-mal-hergucken.html, 31.1.2012.
GetGlue. (2012). *Your app for TV, movies & sports*, getglue.com/about, 11.10.2012.
GetSatisfaction. (2014). *Homepage*, www.getsatisfaction.com/corp/about, 7.2.2014.
Gigya. (2012). *Social privacy survey, consumers want more transparency*, www.blog.gigya.com/announcing-gigya-socialprivacy%E2%84%A2-certification-and-new-consumer-privacy-survey-results, 20.5.2013.
Gillin, P. (2009). *The new influencers: A marketer's guide to the new social media*. Fresno, CA: Quill Driver Books.
Gogoi, P. (2006). *Wal-Mart vs – the Blogosphere (17.10.2006)*, businessweek.com/bwdaily/dnflash/content/oct2006/db20061018_445917.htm?chan=top+news_top+news+index_businessweek+exclusives, 20.4.2010.
Google. (2012). *Our mobile planet: United States, understanding the mobile costumer*, services.google.com/fh/files/blogs/our_mobile_planet_us_en.pdf, 8.1.2014.
Grossman, L. (2006). *Time's person of the year: You (13.12.2006)*, http://www.time.com/time/magazine/article/0,9171,1569514,00.html, 15.1.2014.
Hansell, S. (2008). Zuckerberg's law of information sharing. *The New York Times, 6*(11), 2008.
Harlinghausen, C. S. (2012). *Facebook für Professionals, 3*. Düsseldorf: Social Media Kongress, 27.8.2012.
HCL. (2014). *Customer advisory council*, hcltech.com/customer-advisory-council, 7.2.2014.
Heinrichkeit, L. (2011). Die Fallen des Gefallens. *Frankfurter Allgemeine Sonntagszeitung, 25*, 39.
Heller, P. (2012). Die Masse macht's. *Frankfurter Allgemeine Sonntagszeitung, 73*, November 7, 2012.
Helm, S., & Günter, B. (2006). Kundenwert – eine Einführung in die theoretischen und praktischen Herausforderungen der Bewertung von Kundenbeziehungen. In B. Günter & S. Helm (Eds.), *Kundenwert – Grundlagen – Innovative Konzepte – Praktische Umsetzungen* (3rd ed., pp. 3–38). Gabler: Wiesbaden.
Helm, J., & Jones, R. (2010). Extending the value chain – A conceptual framework for managing the governance of co-created brand equity. *Brand Management, 17*(8), 579–589.
Hofmann, S., Fasse, M., & Postinett, A. (2012). Es geht nur miteinander. *Handelsblatt, 28./29./30.9.2012*, p. 54.
Homburg, C., Kuester, S., & Krohmer, H. (2009). Analysis of the initial strategic situation. In *Marketing management: A contemporary perspective*, New York.
IBM. (2011a). *IBM Global CMO Study Executive Summary*, public.dhe.ibm.com/common/ssi/ecm/en/gbe03433usen/GBE03433USEN.PDF, 25.03.2013.

IBM. (2011b). *From social media to social CRM, what customers want, the first in a two-part series*, public.dhe.ibm.com/common/ssi/ecm/en/gbe03391usen/GBE03391USEN.PDF, 20.5.2013.

IdeaStorm. (2014). *IdeaStorm can help take your idea and turn it into reality*, ideastorm.com, 7.2.2014.

Jenkins, H. (2006). Searching for the origami unicorn – The matrix and transmedia story telling. In H. Jenkins (Ed.), *Convergence culture – Where old and new media collide*. New York: New York University Press.

Kafka, P. (2013). *Cheer up, Facebook investors! Maybe people aren't bailing on Mark Zuckerberg, After all*, allthingsd.com/20130403/cheer-up-facebook-investors-maybe-people-arent-bailing-on-mark-zuckerberg-after-all, 26.5.2013.

Karle, R. (2010). Die Macht der vielen. *absatzwirtschaft*, Sonderheft, 32–38.

Keiningham, T. L., Aksoy, L., Cooil, B., & Andreassen, T. W. (2008). Linking customer loyalty to growth. *MIT Sloan Management Review, 49*(4), 50–57.

Kersch, M. (2013). *Weiblich, ledig, jung sucht ..., Die neue Zielgruppenansprache in der Multi-Channel-Welt*. Vortrag auf dem Dialog-Marketing-Gipfel, Frankfurt, 21.8.2013.

Keylens. (2013). Customer touchpoint management. *Düsseldorf, 2013*.

KLM. (2014). *KLM meet and seat*, klm.com/travel/gb_en/prepare_for_travel/on_board/your_seat_on_board/meet_and_seat.htm, 8.2.2014.

Kreutzer, R. (2009). *Praxisorientiertes Dialog-Marketing, Konzepte – Instrumente – Fallstudien*. Wiesbaden: Gabler.

Kreutzer, R. (2013). *Praxisorientiertes Marketing, Konzepte – Instrumente – Fallbeispiele* (4th ed.). Wiesbaden: Gabler.

Kreutzer, R. (2014). *Praxisorientiertes Online-Marketing, Konzepte – Instrumente – Checklisten* (2nd ed.). Wiesbaden: Gabler.

Kreutzer, R., & Schober. (2010). *Studie zur Interessenten- und Neukundengewinnung, Interpretationen und Thesen*. Ditzingen, Berlin: Ergebnisse.

Lafferty, J. (2012). *Facebook fans: Quality matters more than quantity*, allfacebook.com/napkin-labs-superfans-study_b102500, 18.10.2013.

Lardinois, F. (2012). *Forrester: 84% Of U.S. adults now use the Web daily, 50% own smart phones, tablet ownership doubled to 19% In 2012*, techcrunch.com/2012/12/19/forrester-84-of-u-s-adults-now-use-the-web-daily-50-own-smart phones-tablet-ownership-doubled-to-19-in-2012, 8.1.2014.

Lashinsky, A. (2012). *Inside Apple, Das Erfolgsgeheimnis des wertvollsten und verschwiegensten Unternehmens der*. Welt: Weinheim.

Lecinski, J. (2011). *ZMOT – Winning the zero moment of truth*. Chicago: Google.

Li, C., & Bernoff, J. (2008). *Groundswell: Winning in a world transformed by social technologies*. Boston: Harvard.

Llopis, G. (2012). *Obama and romney are ignoring hispanic voters, HBR Blog Network*, blogs.hbr.org/cs/2012/10/obama_and_romney_are_ignoring.html, 9.4.2013.

Löhr, J. (2014). Facebook? Gefällt nicht mehr so gut. *Frankfurter Allgemeine Zeitung*, 22, 10.4.2014.

Ludowig, K., & Schlautmann, C. (2012). Trügerische Jubelarien. *Handelsblatt*, p. 7, 16./17.18.11.2012.

Mahrdt, N. (2009). *Crossmedia – Werbekampagnen erfolgreich planen und umsetzen*. Wiesbaden: Gabler.

Maier, T. (2012). Buchhandel vor ungewisser Zukunft. *Bonner Generalanzeiger*, p. 9, 10.10.2012.

Marsden, P. (2011). *Interview mit Social Commerce-Experten Paul Marsden, Part 1: When Social media comes to eCommerce*, smartservice-blog.com/2011/12/13/interview-mit-social-commerce-experten-paul-marsden-part1-when-social-media-comes-to-ecommerce, 13.12.2011.

Mathew. (2008). *If the news is important, it will find me*, mathewingram.com/work/2008/03/27/if-the-news-is-important-it-will-find-me/, 27.3.2008.

Mayer-Johanssen, U. (2012). Es geht ums Ganze. *Handelsblatt*, p. 24, 30.8.2012.

Mayer-Uellner, R. (2010). Der Weg ins soziale Netz. *Markenartikel*, pp. 16–18, 7/2010.
McKinsey. (2011). *Big data: The next frontier for innovation, competition, and productivity*, Washington, DC.
McKinsey. (2012). *Turning buzz into gold, How pioneers create value from social media*, mckinsey.de/downloads/publikation/social_media/Social_Media_Brochure_Turning_buzz_into_ gold.pdf, München
Meckel, M. (2011). Weltkurzsichtigkeit. *Der Spiegel*, p. 94f, 3/8/2011.
Mitchell, D. (2013). *The move to Internet TV, by the numbers*, tech.fortune.cnn.com/2013/05/03/the-move-to-Internet-tv-by-the-numbers, 26.5.2013.
n. a. (2012). C&A hängt Facebook-Kleiderbügel auf. *SpiegelOnline*, 8.5.2012.
n. a. (2012). Youtube bringt Spartenfernsehen nach Deutschland. *Bonner Generalanzeiger*, p. 6, November 9, 2012.
n. a. (2012). „Newsweek" erscheint nur noch online. *Financial Times Deutschland*, p. 8, 19.10.2012.
Netzwelt. (2012). *Jedem Otto seine Brigitte: Werbe-Aktion auf Facebook endet überraschend*, netzwelt.de/news/84862-otto-brigitte-werbe-aktion-facebook-endet-ueberraschend.html, 16.10.2012.
Neuland. (2014). *Smart service terminal*, Bergisch Gladbach.
Nielsen. (2011). *State of the Media: The Social Media Report, Q3 2011*, cn.nielsen.com/documents/Nielsen-Social-Media-Report_FINAL_090911.pdf, 26.5.2013.
Nielsen. (2012). *As TV screens grow, so does U.S. DVR usage*, nielsen.com/us/en/newswire/2012/as-tv-screens-grow-so-does-u-s-dvr-usage.html, 26.5.2013.
Nielsen. (2013). *The cross-platform report, Quarter 3, 2012 – US*, de.slideshare.net/Briancrotty/nielsen-cross-platform-report-q3-2012, 23.5.2013.
Oberhuber, N. (2012). Trau bloß keiner Hotelbewertung. *Frankfurter Allgemeine Sonntagszeitung*, p. 45, 21.10.2012.
Oetting, M. (2010). *Ein Überblick: Paid, curated, owned and earned media*, connectedmarketing.de/cm/2010/02/ein-ueberblick-paid-curated-owned-and-earned-media.html, 1.3.2011.
online publishers association, OPA (2012). *A portrait of today' stablet user, wave II*, June 2012, onlinepubs.ehclients.com/images/pdf/MMF-OPA_-_Portrait_of_Tablet_User-Wave_2_-_Jun12_%28Public%29.pdf, 16.04.2013.
PayPal. (2012). *Willkommen im eBay Kaufraum mit PayPal*, Berlin.
Peppers, D. (2012). *The real implications of the 80–20 rule*, linkedin.com/today/post/article/20121002115903-17102372-the-real-implications-of-the-80-20-rule?trk=mp-edit-rr-posts, 27.11.2013.
Peppers, D., & Rogers, M. (2011). *Managing customer relationships, a strategic framework* (2nd ed.). Hoboken: Wiley.
Peppers, D., & Rogers, M. (2012). *Extreme trust, honesty as a competitive advantage*. New York: Penguin.
Peters, T. (1997). *The circle of innovation*. New York: Alfred a Knopf Inc.
Petouhoff, N. I. (2011). Crowd service: Customers helping other customers. In D. Peppers, & M. Rogers (Eds.), *Managing customer relationships, a strategic framework* (pp. 227–234).
Pew Internet & American Life Project. (2013). *Trend data (adults), demographics of Internet users*, pewInternet.org/Trend-Data-%28Adults%29/Whos-Online.aspx, 16.01.2014.
Pohlmann, S. (2012). Trigger statt Tatort. *Tagesspiegel*, 9.10.2012, tagesspiegel.de/medien/neues-fernsehen-trigger-statt-tatort/7229584.html, 10.10.2012.
Porter, M. (1996). What is strategy. *Harvard Business Review* (November–December), 61–78.
Prange, S. (2012). Kooperation statt Konflikt. *Handelsblatt*, p. 52f, 28./29./30.9.2012.
Qmee. (2013). *What happens online in 60 seconds?*, blog.qmee.com/qmee-online-in-60-seconds, 3.2.2014.
Ray, A. (2010). *The ROI of social media marketing: More than dollars and cents*, blogs.forrester.com/augie_ray/10-07-19-roi_social_media_marketing_more_dollars_and_cents, 29.10.2012.
Rechtien, W. (1999). *Angewandte Gruppendynamik* (3rd ed.). Weinheim: Beltz.
Reichheld, F. F. (2003). The number one you need to grow. *Harvard Business Review*, p. 47–54, 12/2003.

Rungg, A. (2012). Kleine Showeinlage. *Handelsblatt*, p. 2, 8.11.2012.
Scheer, U. (2012). Suche Krisenmanager für Shitstorm. *Frankfurter Allgemeine Zeitung*, p. C1, 29./30.9.2012.
Schmidt, G. (2012). *Zukunft ist, wo der Kunde ist – Der effektivste Weg zum Verbraucher*. Vortrag auf dem Dialogmarketing-Gipfel, Frankfurt, 22.8.201.
Schmidt, J., Göbbel, T., & Bchara, J. (2012). *Marktorientierte Unternehmensführung in globalisierten Märkten, BBDO*, batten-company.com/uploads/media/BBDO9_Insights9_4_Marktbearbeitung_in_globalisierten_Märkten.pdf, 28.1.2014.
Searchmetrics. (2013). *Universal Search Results in the Google SERPs – USA 2012*, blog. searchmetrics.com/us/2013/03/28/universal-search-results-in-the-google-serps-usa-2012/, 29.1.2014.
Smith, A. (2012). *17% of cell phone owners do most of their online browsing on their phone, rather than a computer or other device Pew Research Center's Internet & American Life Project*, Washington, DC.
Sohn, G. (2012). *Game over für Facebook und Google? Reboot-Mentalität macht das Netz unberechenbar*, www.absatzwirtschaft.de, 18.2.2012.
Solis, B. (2012a). *The end of business as usual – Rewire the way you work to succeed in the consumer revolution*. Hoboken: Wiley.
Solis, B. (2012b). *Your brand is more important than you think: BrandSTOKE's 9 Criteria for Brand Essence*, briansolis.com, 26.11.2012.
Statista. (2013a). *Anzahl der Internetnutzer in den USA von 2000 bis 2012 (in Millionen), numbers collected from Nielsen and ITU*, de.statista.com/statistik/daten/studie/205251/umfrage/anzahl-der-Internetnutzer-in-den-usa, 8.01.2014.
Statista. (2013b). *Social network usage of U.S. population in 2012, by frequency (in percent)*, statista.com/statistics/191969/social-network-status-updating-frequency-of-us-americans, 26.1.2014.
Stauss, B. (2000). Perspektivenwandel: Vom Produkt-Lebenszyklus zum Kundenbeziehungs-Lebenszyklus. *Thexis, 17*(2), 15–18.
Steimel, B. (2012). *Abschied von AIDA – wie die kreisende Erregung im Netz das Marketing revolutioniert*, smartservice-blog.com/2012/10/04/abschied_von_aida/, 4.10.2012.
Stelzner, M. (2012). *2012 Social media marketing industry report*, socialmediaexaminer.com/social-media-marketing-industry-report-2012, 25.5.2013.
Stüber, J. (2010). *Die Lawine donnert bereits*, welt.de/die-welt/vermischtes/article7297786/Die-Lawine-donnert-bereits.html, 3.6.2010.
Thiel, T. (2012). Was ist denn noch privat? *Frankfurter Allgemeine Zeitung*, p. 27, 22.10.2012.
Trümpler, E., & Neuburger, M. (2012). *„Fuck U!" Schönheits-Preis für Pöbel-Protest*, mopo.de/nachrichten/fotowettbewerb-gewonnen–fuck-u—schoenheits-preis-fuer-poebel-protest,5067140,16941510.html, 22.8.2012.
Vivaldi. (2012a). *Social currency 2012 report*, vivaldipartners.com/vpsocialcurrency/sc2012, 17.12.2012.
Vivaldi. (2012b). *Social Currency 100+*, vivaldipartners.com/vpsocialcurrency, 17.12.2012.
Waissi, G. (2011). *Arizona A&D supply chain*, 129.219.40.44/adsr/Supplychain/SCDefault.aspx., 7.12.2012.
Wiedlich, W. (2010). Datenflut und Datenebbe. *GA-Journal*, p. 1, 6, 6.-7.3.2010.
Winslow, G. (2012). *Nielsen: U.S. Social Media Usage Up 37%, Mobile users now spending 30% of their online time on social media sites*, broadcastingcable.com/article/490668-Nielsen_U_S_Social_Media_Usage_Up_37_.php, 8.2.2014.
Wohlfarth-Bottermann, M. (2013a). *Facebook-Token*, internal paper, Köln.
Wohlfarth-Bottermann, M. (2013b). *GUESS und Tilly's Mobile Commerce Use Cases*, internal paper, Köln.
Woiceshyn, J., & Falkenberg, L. (2008). Value creation in knowledge-based firms: Aligning problems and resources. *Academy of Management Perspectives*, pp. 85–99.

Wüst, C. (2013). „Wenn die Reputation kommt – und geht. In C. Wüst & R. Kreutzer (Eds.), *Wie Journalisten und Analysten die Glaubwürdigkeit und Authentizität der CEO-Kommunikation bewerten*. Wiesbaden: Corporate Reputation Management.
YouTube. (2014). *Statistik*, youtube.com/yt/press/, 25.1.2014.
Zickuhr, K., & Smith, A. (2012). *Digital differences, Pew Research Center's Internet & American Life Project*, Washington, DC.
Zschunke, P. (2012). Die neue Macht des zweiten Bildschirms. *Bonner Generalanzeiger*, p. 3, 5.10.2012.

Further Reading

Andersen, C. (2006). *The long tail*. New York: Hyperion.
Cluetrain. (1999). *Cluetrain Manifesto*, cluetrain.com/cluetrain.pdf, 2.10.2012.
Gladwell, M. (2002). *The tipping point: How little things can make a big difference*. New York: Back Bay Books.
Kreutzer, R. (2009). *Praxisorientiertes Dialog-Marketing, Konzepte – Instrumente – Fallstudien*. Wiesbaden: Gabler.
Kreutzer, R. (2014). *Praxisorientiertes Online-Marketing, Konzepte – Instrumente – Checklisten* (2nd ed.). Wiesbaden: Gabler.
Li, C., & Bernoff, J. (2008). Groundswell: Winning in a world transformed by social technologies. *Harvard*.
Moore, G. A. (2002). *Crossing the chasm: Marketing and selling disruptive products to mainstream customers*. New Work: HarperBusiness Essentials.
Peppers, D., & Rogers, M. (2011). *Managing customer relationships, a strategic framework* (2nd ed.). Hoboken: Wiley.
Peppers, D., & Rogers, M. (2012). *Extreme trust, honesty as a competitive advantage*. New York: Penguin.
Qualman, E. (2010). *Socialnomics: How social media transforms the way we live and do business*. New York: Wiley.
Solis, B. (2010). Engage: The complete guide for brands and businesses to build, cultivate, and measure success in the new web.
Solis, B. (2012). *The end of business as usual – Rewire the way you work to succeed in the consumer revolution*. Hoboken: Wiley.
Solis, B. (2013). *What's the future of business? Changing the way businesses create experiences*. Hoboken: Wiley.

GPSR Compliance

The European Union's (EU) General Product Safety Regulation (GPSR) is a set of rules that requires consumer products to be safe and our obligations to ensure this.

If you have any concerns about our products, you can contact us on

ProductSafety@springernature.com

In case Publisher is established outside the EU, the EU authorized representative is:

Springer Nature Customer Service Center GmbH
Europaplatz 3
69115 Heidelberg, Germany

www.ingramcontent.com/pod-product-compliance
Lightning Source LLC
LaVergne TN
LVHW010338260326
834688LV00036B/777